Actuaries' Survival Guide

Second Edition

Actuaries' Survival Guide

How to Succeed in One of the Most Desirable Professions

Second Edition

Fred E. Szabo, PhD

Department of Mathematics and Statistics
Concordia University

AMSTERDAM • BOSTON • HEIDELBERG • LONDON
NEW YORK • OXFORD • PARIS • SAN DIEGO
SAN FRANCISCO • SINGAPORE • SYDNEY • TOKYO
Academic Press is an imprint of Elsevier

Academic Press is an imprint of Elsevier
225 Wyman Street, Waltham, MA 02451, USA
525 B Street, Suite 1900, San Diego, CA 92101-4495, USA
The Boulevard, Langford Lane, Kidlington, Oxford OX5 1GB, UK

Library of Congress Cataloging-in-Publication Data
Application submitted.

British Library Cataloguing-in-Publication Data
A catalogue record for this book is available from the British Library.

ISBN: 978-0-12-386943-2

For information on all Academic Press publications
visit our website: *http://store.elsevier.com*

Typeset by: diacriTech, India

Printed in the United States of America
12 13 14 15   9 8 7 6 5 4 3 2 1

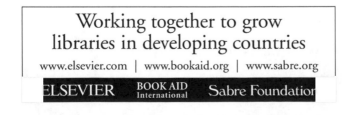

Prediction is very difficult, especially about the future.
Niels Bohr, Physicist and Nobel Laureate

CONTENTS

ACKNOWLEDGMENTS

This second edition of the Actuaries' Survival Guide is the brainchild of Patricia Osborn, an Acquisitions Editor at Elsevier Academic Press. As she examined the continuing success of the guide over close to ten years and in view of the changes that have taken place in the actuarial profession during that time, the thought occurred to her that it might be appropriate to update some of the material in this guide.

Initially it seemed relatively straightforward to update the guide with the help of several enthusiastic reviewers: Claire Bilodeau, Associate Professor, Director, Undergraduate Program, École d'actuariat, Université Laval; Professor David Hudak, Director, Actuarial Science Program, Robert Morris University; Paul H. Johnson, Jr., Assistant Professor, Department of Mathematics, University of Illinois at Urbana-Champaign; and Toby White, Assistant Professor of Actuarial Science and Finance, Drake University. While they will recognize many of their suggestions implemented in this book, the final version of this guide is entirely my own responsibility.

What made the updating of this guide difficult as I began working on it is the fact that the first edition is based on a survey of actuarial students and professionals that depends critically on the actuarial examination structure that existed at the time when I conducted the survey. Since then, the actuarial examinations have been restructured and often renamed. On the other hand, the mathematics of actuarial science has changed very little and the type of examination questions used to illustrate the mathematics in the first edition of this book are as relevant today as they were in the past. I therefore decided to keep intact the references to specific actuarial courses that predate today's organization of the material by SOA, the

Society of Actuaries, and by CAS, the Casualty Actuarial Society. Various tables included in the text serve as roadmaps to the new structure. What has remained relatively constant is the challenge of becoming an accredited actuary by succeeding in the actuarial examinations. However, the process has become more transparent and perhaps easier because of the wealth of information and suggestions in print and online. On the other hand, the amount of material available today to aspiring actuaries is simply overwhelming. Fortunately, social media such as **Facebook** and **LinkedIn** are helping to personalize and humanize the process. For this reason, Appendices A and B are built around information easily available not only on company websites, but also on **Facebook**, **LinkedIn**, and, in the case of consulting firms, even **Twitter**.

As your try to come to grips with the challenge of comparing the various actuarial career options and the available pathways to succeed, try to keep in mind that no matter which route you choose, the actuarial world is built around four types of knowledge: core knowledge common to all options, specialized knowledge depending on the field, professional knowledge, and experiential knowledge. If you love math and think that you might enjoy mixing it with business, law, and accounting, for example, you should seriously consider becoming an actuary. The personal stories on which this guide is based will illustrate how some of your predecessors managed to get there.

I would like to thank the following actuaries, actuarial students, mathematicians, economists, consultants, and career experts for having participated in the original design and completion of the actuarial survey on which the hands-on material in the book is based:

Jonathan Bilbul, Marie-Andrée Boucher, David Campbell, Steve Cohen, François Dauphin, Karine Desruisseaux, Pierre Dionne, Norman Dreger, Jean Drouin, Julie Duchèsne, Louis Durocher, Amélie Girard, Philippe Gosselin, Karine Julien, David Laskey, Dany Lemay, Erik Levy, Jean-Grégoire Morand, Paul Morrison, Lambert Morvan, Dylan Moser, Céline Ng Tong, Marc Parisien, Karlene Parker, Caroline Piché, Étienne Plante-Dubé, Elisabeth Prince, Graham Rogers, Martin Rondeau, Siobhain Sisk, Mariane Takahashi, Véronique Tanguay, and Ghislaine Yelle. The information that they have supplied in the survey or by direct communication, and that supplied by others, is reproduced in this book in anonymous and sometimes paraphrased form because the survey was designed to guarantee the confidentiality of the answers. The survey was completed on the understanding that the opinions expressed are personal and should not be construed as representing the views of the companies where many of the respondents are employed.

I would also like to acknowledge the contributions of the reviewers of the first edition of this book. They include Dale Borowiak (University of Akron), Bruce Edwards (University of Florida), Louis Friedler (Arcadia University), José Garrido (Concordia University), Brian Hearsey (Lebanon Valley College), Jon Kane (University of West Washington), Stuart Klugman (Drake University), Jean Lemaire (University of Pennsylvania), Murray Lieb (New Jersey Institute of

Technology), Vania Mascioni (Western Washington University), Charles Moore (Kansas State University), Kent Morrison (California Polytechnic State University), Walter Peigorsch (University of South Carolina), Gabor Szekeley (Bowling Green State University), Charles Vinsonhaler (University of Hawaii), and Bostwick Wyman (Ohio State University), as well several anonymous referees who provided guidance with the design of the project. I would like to thank all of them for their constructive comments. They will recognize traces of their ideas throughout the text.

The Society of Actuaries has granted me permission to build Chapter 2 around sample questions and answers from its examinations, the Casualty Actuarial Society has granted me permission to include the results of their survey on CAS professional skills, and the International Actuarial Association has permitted me to include its list of competency areas of actuaries. I hereby express my sincere thanks to them.

Others have provided direct information in other forms and I am grateful to them. They include Michelle Aspery (Institute of Actuaries of Australia), Malcolm Campbell (COO Skandia Offshore Business), Maria da Luz Fialho (Portuguese Institute of Actuaries), Peter Diethelm (Association Suisse des Actuaires), Wim Els (Actuarial Society of South Africa), Yves Guérard (International Actuarial Association), Caroline Henderson-Brown (The Actuarial Profession), Betty-Joe Hill (Royal & SunAlliance), Curtis E. Huntington (University of Michigan), Liyaquat Khan (Actuarial Society of India), Pat Kum (Actuarial Society of Hong Kong), Dr. Eduardo Melinsky (University of Buenos Aires), Dr. Mario Perelman (Argentinian Institute of Actuaries), Dr. Jukka Rantala (University of Helsinki), Loredana Rocchi (Italian Institute of Actuaries), Deborah R. Rose (Institute and Faculty of Actuaries), Dr. Rafael Moreno Ruiz (Universidad de Málaga), Nicole Séguin (International Actuarial Association), Martha Sikaras (Society of Actuaries), Elizabeth Smith (Casualty Actuarial Society), Stuart Szabo (Global Corporate Finance, Deutsche Bank), Klaus Wegenkittl (Union Versicherungs-Aktiengesellschaft), Karin Wohlgemuth (Zurich Financial Services), Yew Khuen Yoon (Actuarial Society of Malaysia), Masaaki Yoshimura (Institute of Actuaries of Japan), and Aleshia Zionce (Society of Actuaries).

I would also like to express my sincere appreciation and gratitude to Dr. Harald Proppe, my colleague, and to Eric Hortop, my former student, for having spent many hours reading the manuscript and suggesting corrections and improvements. Eric has been particularly helpful with the updating of some of the factual information on which this second edition of the guide is based. As the manuscript reached completion, I received an unexpected offer of help from Jean-Grégoire Morand, senior actuarial consultant at Normandin Beaudry, to help me proofread the manuscript before submitting it to Elsevier for publication. I jumped at the opportunity to have Greg take an unbiased look at my work. The quality of this guide owes much to his keen eye and thoughtful suggestions. Thank you, Greg.

A special thanks is reserved for Kathryn Morrissey (Elsevier Academic Press), my Editorial Project Manager, who guided me joyfully and effectively through the production of the second edition of this book.

The anonymous survey was produced and evaluated with the creative help and guidance of Maggie Lattuca, formerly of the Concordia University Instructional and Information Technology Services Department and now the Manager of Educational Technologies at McGill University, Montreal, using *Respondus* and *WebCT*.

Both editions of this book were written in LATEX using *Scientific WorkPlace*, a mathematician's dream come true. The writing of this second edition was made even more enjoyable because of many recent enhancements to *Scientific Work-Place* and the fact the latest version also runs on Macintosh computers. I would like to thank Barry MacKichan and his team for continuing to produce and improve this unique scientific writing tool.

Fred E. Szabo
23 October 2011

INTRODUCTION

You are reading this book because you are thinking about the future. What would you like to do with your life? What career would allow you to fulfill your dreams of success? If you like mathematics, your choices have just become simpler. Consider becoming an actuary.

Almost ten years ago, I wrote the first edition of the *Actuaries' Survival Guide* to explain what actuaries are, what they do, and where they do it. The publishers, Elsevier Academic Press, encouraged me to prepare a second edition of this little book for two reasons: the book has been remarkably well received by its readers and the required changes could be made without destroying the coherent and conversational flow of the narrative that was so effective in the first edition.

Positive comments on the first edition of this book by its readers have kept me going as I set out to revise the guide:

- "Anyone linked to this profession **must** have this book." (Michael)

- "I am a research mathematician who has often wondered what my life would be like today, had I chosen to become an actuary. Now I know. Dr. Szabo's insider profiles of working actuaries are entertaining, informative, and comprehensive. I have enjoyed the accounts of the day-to-day activities of actuaries at their various levels of seniority, and have found it intriguing to learn how some of the underlying mathematical ideas and techniques can be discussed in a book of this type without assuming advanced prior knowledge. I highly recommend this guide to anyone interested in an actuarial career." (Chantel)

- "I have been searching with my students for a good book explaining [the] challenges of the actuaries' job and the way to do it. I could not find anything that would guide my students and help them in making an informed decision. Finally, I got the Actuaries' Survival Guide by Fred Szabo.... The content of the book is organized in a very logical way. It is as complete... as possible. We learn [about] the actuarial job, what skills we need for it, types of accreditation, what education we must go through, [and] how to start and continue an actuarial professional career.... In conclusion I would like to say this is a wonderful book written in a professional way for the people who wish to be professionals in one of the most difficult professions." (Mirek)

Several things have changed since I wrote the first edition of this book:

- *Distinct syllabi.* The examination structures of the Society of Actuaries and the Casualty Actuarial Society of America have been modified. They now have more focused paths to success.

- *Online examinations.* The revised examination processes now include computer-based online examinations known as *CBTs.*

- *Social networking.* Actuarial companies have come and gone, and the network of actuaries, both for job searches and advances in employment, have changed as a result of the explosion of social networking.

- *New markets.* The marketplace has changed. Steady progress towards a global knowledge-based economy has opened up new needs for actuaries and new employment opportunities.

- *Global expansion.* The need for actuarial services has grown: changes in health insurance in the United States, for example, require new actuarial models.

- *Internet growth.* As in many other fields, the Internet has changed the way actuaries study and interact and the way actuarial work is done.

In spite of these changes, the principles, skills, and techniques on which the actuarial profession is based and depends have remained stable. Although the actuarial accreditation processes were revised in 2003 and the SOA and CAS syllabi were realigned, the methods for succeeding in the associated examinations, as described in the first edition of this guide, have essentially remained unchanged. The responses to the survey and the references to pre-2003 courses are valid and helpful. To distinguish these courses from those offered today we add the tags [2002] or [2003] to courses that were offered on or before the time the survey was conducted.

Since the courses and examinations required to become an accredited actuary are subject to ongoing changes, references to courses and examinations in this book are meant as illustrations only. You should always check with the societies' websites for the up-to-date course outlines and all matters related to actual examinations.

What Is an Actuary?

You may have heard several descriptions of actuaries and their work. Are they mathematicians, statisticians, economists, investment bankers, legal experts, accountants, or business experts?

I will show you that an actuarial career involves elements of all of these professions and more. I will try to open the rich mosaic of actuarial life for you. As such, I would like this book to be more than a career guide. I would like it to be your career companion on the road to professional success.

When discussing this project with a senior actuary, I was told that writing a book about actuaries is like trying to shoot at a moving target. The book needs to be updated as soon as it is written. This fact made the project an even greater challenge and more exciting. It made me realize that my job was to concentrate on the big picture, the ideas and scenarios that unite this changing and dynamic world and give it permanence. The book in your hands is the result.

This is a hands-on book. For more than fifteen years, I have been associated with actuarial students as the director of an actuarial cooperative program at Concordia University. Before writing this book, I consulted over 100 of my former students and their employers about what kind of introduction to their profession they would have liked to have had when they were making their career choices. Their answers form the background to this book.

The Syllabus

As you explore the different features of this guide, you will appreciate its global perspective. You are contemplating becoming an actuary. It therefore makes sense to consider all of your options so you can decide for yourself why you might want to do so and assess how to succeed in one of the most desirable professions.

You have probably heard that you need to study long and hard to become an accredited actuary. But what do you need to study and how does the subject matter you are required to master in North America, for example, compare with that required elsewhere? By comparing different syllabi you will quickly get an idea of what is involved. You can use this book to compare several actuarial syllabi:

1. The syllabus of the International Actuarial Association
2. The syllabus of the *Groupe Consultatif Actuariel Européen*
3. The 2002 syllabus of the Society of Actuaries
4. The 2011 syllabus of the Society of Actuaries

5. The 2002/2003 syllabus of the Casualty Actuarial Society
6. The 2003 syllabus of the Casualty Actuarial Society

The syllabi of the Society of Actuaries and Casualty Actuarial Society have two entries: those up to 2002 and 2003, and those of 2011, for several reasons:

- The most appreciated aspects of the first edition of this guide were the question-and-answer sections about the actuarial profession. Although many of the cited answers refer to the 2002 and 2003 syllabi of the Society of Actuaries and Casualty Actuarial Society, they are as meaningful today as they were ten years ago.

- If you compare the mathematical topics in the 2002 and 2003 syllabi with those from 2011, you will find that mathematically, not much has changed. The topics have been reorganized and online methods of learning and teaching have helped facilitate delivery of the information. The emphasis on professionalism and professional development, which is ever present in the 2011 syllabi, did not form part of the survey. Hence no change was required in the revision of this guide.

  When you look at the sample examination questions and their solutions, you are of course not meant to know it all. You are encouraged, however, to relate some of the topics to your complementary course and get a sense of the academic direction your studies should take.

- In addition, you can use the "Actuaries around the World" section to compare these syllabi with the accreditation systems in several other countries and, by perusing the reciprocity agreements between different countries described in Appendix C, you can develop a feel for the remarkable international portability of actuarial credentials, unmatched by almost any other profession.

The international nature of the actuarial profession is again illustrated in Appendices A and B where the national and international locations of many of the listed companies are included to allow you to dream of working in some of your favourite cities, states, provinces, and countries around the globe.

You may also wonder why the basic actuarial symbols are illustrated in this guide. One obvious reason is that some of them are used in the sample examination questions in the book. But a deeper reason is the universality of the actuarial language, which adds another dimension to the portability of actuarial knowledge, experience, and techniques across national borders. Once again we recognize actuaries as a global community of experts.

Q&A

Among the key features of this book are the answers provided by over 50 actuaries and actuarial students in an electronic survey about the actuarial profession, sent to over 150 experts. The submitted answers are included in the book, with minimal editing to preserve their flavor and the spontaneity of the replies.

In assembling the material for this book, I asked some of my former students and some of their employers for comments on my plan. Here is what they had to say:

Q **Which topics in this book do you consider to be the most important and why?**

Answer   Non-actuarial opportunities for students who enter an actuarial program at university.

Answer   Understanding the full range of career options enables a better career choice with increased odds of job satisfaction and high-profile success.

Answer   Technical skills (more important than interpersonal skills), internships (the best way to learn about the profession and find a full-time job), career profiles (since there are more profiles than people can imagine), actuarial recruiting (not always well known).

Answer   I would say, equally, employers and careers because we don't learn that in school. Sometimes, teachers have not worked in companies, so they may not be familiar with this information. It can be a concern at any level, from high school to university.

Answer   The chapter about the different career possibilities. I think this is important because as long as we are not working in a particular field, it is really hard to have an idea of what it is about. A clear description of these career opportunities would be really helpful for choosing both an internship and a permanent job.

Answer   The skills. A lot of people do not know if an actuarial career is for them. Other important topics are the profession and the industry. I believe that the actuarial profession is not for everyone and that one has to love it to be happy and successful in it. Knowing what an actuarial career is all about is important before embarking on it. Other important topics are the professional courses; becoming an actuary is a lot of work and being aware of the studying that it required is important before taking the decision to opt for an actuarial career.

Answer   Real world applications of exam material, just so that students taking the exams feel that what they are learning is actually useful.

Answer   The differences between different actuarial fields. For example, risk management is an area of interest to me. Another topic of interest to me is nontraditional areas of work.

Answer You need to know what is required to be a *good* actuary and whether you are suited for this career. Moreover, the choice of SOA [pensions, health, finance, consulting, etc.] or CAS [property and casualty insurance, etc.] is extremely important. I was in the SOA stream when I thought of changing fields. Once I had the opportunity to work with a P/C [property and casualty] insurer, I saw the light and decided to switch to P/C (and I love it). I must say that school did not help me in making this right decision. Not enough information was given to students about the CAS option.

Answer The part concerning the SOA and CAS courses since, in my opinion, they are the most important aspect of an actuarial education.

Answer The chapter dealing with SOA and CAS career choices and a list of leading employers interest me the most.

Answer An actuarial background is a great asset for many more jobs than people might think. I am sure that in the near future only a small percentage of students graduating in actuarial mathematics will actually do actuarial work. They might not be typical actuaries but can easily become, given the necessary personal skills, great leaders in different areas of the business world. An actuarial mathematics background opens the door to a vast world of opportunities in the marketplace.

Answer It would be interesting to have a listing of companies with the type of jobs they are offering.

Note: Throughout this book, we use the acronyms SOA and CAS both as names for the *Society of Actuaries* and the *Casualty Actuarial Society* and as designations for *careers* for which an Associateship or Fellowship in these or similar societies is normally required.

Several responses refer to the salary you can expect at various stages of your career. The cited figures refer to the salaries at the time the survey was conducted. These salaries have obviously changed over time. Salaries today are about $20-25\%$ higher today than they were at the time the survey was conducted. D. W. Simpson, the actuarial recruiting firm, regularly publishes an in-house list of actuarial salaries for various employment groups on their website at www.dwsimpson.com that is more up-to-date. Comparing the new ranges with those listed by the respondents provides an interesting basis for comparison and might help you as you speculate about what your salary range may be as you advance in your actuarial career.

The Casualty Actuarial Society also publishes a salary table for property and casualty compensation at www.actuarysalarysurvey.com.

According to the Wolfram Research searchable database of public numerical data, available at www.wolfram.com, the recent salaries of actuaries in the United States were as follows:

- The average salary of an actuary in 2008 was US$84,810.

- Between 2001 and 2008, the average salary of an actuary rose from approximately US$70,000 to US$85,000.

- In the 50% range, the average salary of an actuary was in the US$62,000 to US$119,110 range.

- In the 80% range, the average salary of an actuary was in the US$49,150 to US$160,780 range.

To avoid any possible confusion between the salaries in the question-and-answer sections of the book and the current salaries, I have added the tag [2002] to all salaries mentioned by the respondents to the survey.

The Examinations

Another useful feature of the book is the inclusion of sample questions from joint Society of Actuaries (SOA) and Casualty Actuarial Society (CAS) examinations. The letters and the names of the SOA and CAS examinations referred to in the questions-and-answer sections have changed since 2003. However, the content of these examinations has not changed substantially. I have therefore kept the *look-and-feel* format of the provided answers and kept editing to a minimum.

By browsing through these sections, you will get an idea of what the examination questions *look* like and get a *feel* for what the answers should be. Although the form and content of these examinations will continue to change over time, the ideas and techniques presented in the given examples will give you an idea of the mind set of professional actuaries.

Chances are that neither the questions nor their answers will make sense to you at this time. But by perusing their content and looking at the form of their answers, you will get a sense of what lies ahead.

In order to pass the SOA and CAS examinations, you must use some of the techniques and study tools discussed in Chapter 2. You will find a list of appropriate references on the websites listed in Appendix D. In the case of SOA Courses 5 [2002] through 8 [2002], and CAS Courses 5 [2002] through 8 [2002], you will only find summaries of the course descriptions. They will give you an idea of the content of these courses. You will quickly realize that the courses are based on ideas and techniques related to actuarial work experience. Students who manage to pass the first four foundation examinations usually learn from their colleagues what to expect in these advanced courses and how to prepare for their examinations.

Global Employers

A further aspect of this book that you will find useful is a list of typical employers. The list is incomplete because there are thousands of public and private companies as well as government agencies employing actuaries. It is based on personal contacts and suggestions received from respondents to the survey. As presented, the list is meant as a starting point for your personal research into actuarial employment. The details provided about different companies differ from employer to employer. The intention is to highlight different aspects of employment rather than giving an encyclopedic description of employment at particular companies. You can easily complete the sketches by consulting the cited websites. Most of the included employers have been chosen because of their global reach. They typify the extraordinary opportunities that exist worldwide for actuarial employment.

Online Resources

Several online resources have been created to highlight the fact that the actuarial profession is essentially a close-knit family of professionals that cares deeply about the success of its members. This is probably one of the reasons why ongoing surveys have shown actuaries belong to "one of the most desirable professions."

The website entitled *Be An Actuary* (see www.beanactuary.org) is definitely required reading for aspiring actuaries at all levels: high school, college, and university. The site contains answers for students, parents, educations, counselors, and employers.

Not surprisingly, social media has also entered the fray. The Facebook page at www.facebook.com/Actuarial, maintained by The Actuarial Profession, an industry organization based in the United Kingdom, is in particular a helpful source of information and support.

Other media, such as Twitter, also have a good deal of interesting and relevant information about actuaries and their careers (see, for example, http://twitter.com/TheActuaryjobs). Individuals and companies advertise positions, announce seminars and meetings, and create bonds between professionals and students.

In addition, online resources such as the Actuarial Bookstore (see www.actuarial-bookstore.com) have publications with information on various topics, including how to prepare for your first actuarial examination, beginning your actuarial career, overviews of skills and exams, actuarial internships, full-time positions, and employer actuarial student programs.

As you begin to browse the Internet for information on actuaries and the actuarial profession, you will come across more source material than you can handle. However, the following resources, listed in the Bibliography, among many others, may be of particular interest: The Actuarial Career Planner [1], the CAS Survey of Professional Skills [8], and the CAS Syllabus [9]. Some time ago, I also wrote the Graduate Student Success Guide [19], which discusses planning and study strategies that apply not only to graduate studies, but to the actuarial profession as well, and you may find it helpful. If you compare the first edition of this Guide [21] with this new version, you will get a sense of the rapid pace at which the actuarial profession has evolved, has become more global, and will no doubt continue to change as the world becomes a smaller and smaller global village. The publication entitled *The Future Actuary* [22] should therefore definitely be on your reading list.

Another excellent source for preparing for the first four examinations of the Society of Actuaries and the Casualty Actuarial Society is the quarterly newsletter *The Future Actuary* (see www.soa.org/news-and-publications), which provides readers with insight into topics such as career development and study tips focusing on the first four jointly administered actuarial examinations.

Resources in Print

Much of the material integrated in this book is in the public domain and is available in bits and pieces through a multiplicity of published sources. However, it is widely scattered and incoherent and, as such, appears disjointed and overwhelmingly complex. One of the objectives of this project was to analyze the available information and build a coherent picture.

The sources of the quoted material are always clear from the context and are often explicitly mentioned. When required, italics are used to identify passages quoted verbatim. However, the world does not stand still and the source of the quoted material may have changed over time. This is especially important to keep in mind in the case of actuarial examinations and the rules that govern them. You should always consult the up-to-date websites or latest editions of published material for revisions of content, rules, and requirements. The same applies to things such a reciprocity agreements and other legal matters. The guide is primarily meant as a motivating roadmap through the maze of the actuarial profession, enriched with the enlightening, constructive, and supporting comments from working actuaries, actuarial employers, and actuarial students.

▶ Chapter 1

ACTUARIAL CAREERS

1.1 Career Options

The word *actuary* comes from the Latin word *actuarius*, which referred to short-hand writers in the days when things like typewriters and computers hadn't even been thought of. Today, actuaries work for insurance companies, consulting firms, government departments, financial institutions, and other agencies. They provide crucial predictive data upon which major business decisions are based. True to their historical roots, actuaries still use a rather extensive shorthand for many of the special mathematical functions required for this work (see [5], pages 687–691; [18], pages 123–131; Appendix H [Note: All references are listed in Appendix I.]). The sample questions and answers for Courses 2 [2002] and 3 [2002] discussed in Chapter 2 illustrate some of the currently used actuarial symbols listed in Appendix H. The symbols are an amazingly rich combination of right and left sub-scripts and superscripts, attached to designated upper- and lowercase Roman and Greek letters. They provide for a unique and common actuarial language around the world.

Actuarial science is an exciting, always-changing profession, based on fields such as mathematics, probability and statistics, economics, finance, law, and business. Most actuaries require knowledge and understanding of all of these fields and more. To ensure that this is really the case, all actuaries must pass special examinations before being recognized as members of the profession. To perform their duties effectively, actuaries must also keep abreast of economic and social trends, as well as stay up-to-date on legislation governing areas such as finance, business, health care, and insurance.

No doubt you have heard about the actuarial examinations you need to pass to become an *Associate* or *Fellow* of one of the actuarial societies. Often full-time employees in actuarial firms who are still engaged in the examination-writing process are distinguished from Associates and Fellows by being referred to as *Student Actuaries*. The efforts required to succeed in these examinations are in many ways analogous to those required to become a doctor, lawyer, or other high-ranking professional. So are the rewards. For several years now, the Jobs Rated Almanac has considered an actuarial career to be one of the most desirable professions in America (see [15]).

Actuaries are experts in the assessment and management of risk. Traditionally, the risks managed by them have been insurance and pension funding risks, although the management of business risks is also among the responsibilities of insurance actuaries. So is the insurance of insurance, known as *reinsurance*. Moreover, many actuaries are now also managing asset-related risks in merchant banks and consulting firms. This bodes well for the long-term future of the profession, since risks of all kinds will always be with us. The global financial crises plaguing the modern world highlight the importance of being able to measure financial risks and being able to develop plans for managing them. However, as you will see later on in this book, the day-to-day activities of an actuary depend very much on the sector of the financial services industry where the actuary works.

Actuaries are often chosen to be general managers in insurance companies. This is because upper management and boards of directors have a high regard for the knowledge and skills of actuaries, and because the need of a company to maintain its financial integrity makes an actuary's numerical skills invaluable.

Actuarial Terms, Acronyms, and Definitions

As you read on, you will quickly discover that actuarial science is full of technical terms, acronyms, and definitions. This book is not the place for explaining them in detail, because the definitions involved are readily available in textbooks and on the Internet. The main objective of this book is to introduce you to the career opportunities that exist in the actuarial world and to sketch for you the steps required to enter that world. For this reason, most of the technical material in the book is provided only in illustrative and summary form. Consider it a detailed roadmap to the relevant topics in mathematics, business, and statistics. It is merely meant to help you identify the range of knowledge involved in actuarial work. The study of the mentioned topics requires specialized sources and tools. The reference section at the end of the book provides you with the necessary pointers.

Actuaries can be grouped in different ways. As their functions change in response to changes in the world around us, the distinctions become less sharp. However, the following categories of employment will give you an initial idea.

Valuation Actuaries

Reserves are important to the long-term financial health of a company. Because insurance companies are dealing with events that are uncertain in time and amount, they must put aside what they consider to be the most likely amount of money they will need to pay for future claims and expenses, and then put aside a little more, just in case. The role of *valuation actuaries* is to determine the appropriate "just a little more" and to validate the expected number of claims and amounts to be taken into account when setting the price of insurance. Valuation actuaries also certify required reserves to government agencies.

Pricing Actuaries

Pricing actuaries are responsible for determining how much money a company is likely to make on a product. A product can be life insurance, which pays an agreed-upon sum to your beneficiary when you die, an annuity, which pays an agreed-upon sum every month as long as you live, or some form of health insurance, which covers the costs of medical care not paid for by a government plan, for example, dental and drug expenses. Pricing actuaries use the same assumptions as valuation actuaries when calculating the price of insurance to guarantee consistency and ensure that when valuation actuaries believe that they are adding a little extra to the reserves, they are really doing so. We might say that pricing actuaries add *pricing recipes* to insurance products to make them profitable.

Consulting Actuaries

Consulting actuaries spend a good deal of their time advising on benefit programs and defined benefit pension plans. These are trusts set up to fund tax-assisted retirement benefits at a rate spelled out in a legally certified document.

In the United States, senior consulting actuaries are usually members of the Conference of Consulting Actuaries (CCA). To become a Member of the CCA, candidates must have completed a minimum of 12 years of responsible actuarial work, defined as *work that requires knowledge and skill in solving actuarial problems.* In the United States, they must be a Fellow or Associate of the Society of Actuaries or the Casualty Actuarial Society; be enrolled with the Joint Board for the Enrollment of Actuaries (EA), thus having acquired the title of *Enrolled Actuary*; or be a Member of the American Academy of Actuaries (MAAA).

In the United States, an actuary must be an Enrolled Actuary to have signing authority. The Employee Retirement Income Security Act of 1974 specifies that they must therefore have participated in determining *that the methods and assumptions adopted in the procedures followed in actuarial services are appropriate in the light of all pertinent circumstances.* They must also demonstrate *a thorough understanding of the principles and alternatives involved in such actuarial services.* Their actuarial experience must include involvement in *the valuation of the liabilities of pension plans, wherein the performance of such valuations requires the application of principles of life contingencies and compound interest in the determination, under one or more standard actuarial cost methods, of such of the following as may be appropriate in the particular case: normal cost, accrued liability, payment required to amortize a liability or other amount over a period of time, and actuarial gain or loss.*

In the United Kingdom, Canada, and certain other countries, *Appointed Actuaries* play a role analogous to that of Enrolled Actuaries in the United States.

Pension Actuaries

Pension actuaries look at the ages and salaries of all members of a pension plan, and project how much each would receive at retirement on average, given consideration of various future events, including that some members will terminate before retirement and some will get salary increases. Then they look at the assets the pension plan has invested and determine, based on these two analyses, how much the plan's sponsor (usually an employer) needs to contribute to the plan each year. The pension actuary certifies that the contributions needed to fund the plan are adequate and qualify for a tax deduction for the sponsor.

Pension laws and pension regulations are country-specific or even state-specific (United States) or province-specific, as is the case in Canada. This is the one area in which the global mobility of actuaries is somewhat restricted. Special examinations must be passed in the country of employment to be a pension actuary. In the United States, pension actuaries must be Enrolled Actuaries to be eligible to perform government-related pension fund audits. Enrolled actuaries are also employed in the human resource departments of large companies.

Senior pension actuaries in the United States are usually also Fellows of the American Society of Pension Professionals & Actuaries (ASPPA), a designation that is awarded only after successful completion of a series of professional examinations. The basic examinations are those required to become an Enrolled Actuary, together with three additional ASPPA examinations. A Fellow of the American Society of Pension Professionals & Actuaries must also be a Fellow or Associate of one of the following societies: the Society of Actuaries, the Casualty Actuarial Society, the Canadian Institute of Actuaries, the Institute and Faculty of Actuaries, or be a Member of the American Academy of Actuaries (AAA), the Asociacion Mexicana de Actuarios Consultores, the Asociacion Mexicana de Actuarios, or the Colegio Nacional de Actuarios.

Financial Actuaries

As the worlds of banking, insurance, and finance become more entwined, a new breed of actuary is emerging, known as a *financial actuary*. More and more, financial actuaries tend to become "risk actuaries," employed to model types of financial risks. An advertisement for a senior financial actuary on the Internet describes one of the novel roles of actuaries in business. A company was looking for a senior financial actuary whose responsibilities included *developing, analyzing, and testing models of Internet credit card processing systems including product pricing, positioning, and consumer credit, in order to minimize risk and improve return on investment. You will communicate assumptions, results, and alternatives to staff and provide guidance in systems reengineering.* A suitable candidate was expected to have at least a Bachelor's degree in actuarial science, finance, mathematics, or a related field and be an Associate Actuary. In addition to appropriate experience, the candidate was expected to be an effective communicator, and creative thinking skills were essential. The company was looking for a self-starter with a strong statistical background and proven expertise in modeling techniques. Moreover, knowledge of the financial and management needs of an Internet real-time credit card processing company was expected.

Risk Managers

In May 2011, the Institute and Faculty of Actuaries in the UK published a paper entitled *Actuaries in Risk Management* (see [11]). In its preamble, the paper explains that *with their deep expertise in risk, actuaries are widening their influence with careers moving beyond financial services into newer areas such as healthcare and energy. The actuarial qualification provides a foundation that is both broad and technically detailed, helping to equip actuaries to play to their personal strengths to expand their universe of opportunities. It is, therefore, no surprise to find actuaries aiming to be to the fore in the emerging fields of risk management, including the development of risk-based approach to managing an enterprise.* The paper discusses professional trends in the actuarial profession and places considerable emphasis on the evolving role of actuaries as risk managers. The paper is predated by a study by the Casualty Actuaries Society published in 2003 that assessed the role of actuaries in ERM (enterprise risk management) and created a framework for its technical foundations and actual practice. ERM is now a key component of the CAS syllabus (see Chapter 2). Another step forward occurred four years later in 2007 when the Society of Actuaries created the *Chartered Enterprise Risk Analyst* (CERA) designation. The CERA designation of actuaries, also discussed in Chapter 2, has become an enticing career opportunity for actuaries around the world.

What Does It Take to Become an Actuary?

Skills needed to become an actuary certainly include mathematical ability, knowledge of and comfort with computers and computer modeling systems, and the ability to communicate complex topics in terms that customers can understand.

Formally, most actuarial positions require that you are at least an Associate of the Society of Actuaries, the Casualty Actuarial Society, or the Canadian Institute of Actuaries, or have equivalent standing in an actuarial society of another country. If you are in a position that requires you to certify actuarial valuations and reports, you must usually be a Fellow of these societies.

Many actuaries in the United States are also members of the American Academy of Actuaries (see Appendix D), the public policy, communications, and professionalism organization for all actuaries in the United States.

As Section 1.13 shows, actuaries in different countries belong to a wide variety of national and international professional organizations that define and direct the future of the profession. At the international level, the International Association of Actuaries (see Appendix D) plays a central role in coordinating and advancing global actuarial interests.

Rapid and profound socioeconomic changes around the globe bring with them specific problems that need to be solved and financial risks that need to be understood and managed by actuaries. Situations arise that require that the actuaries attending to these problems are exceptionally creative and able "to think outside the box." The growing emphasis on on-the-job experience and continued professional development by actuarial societies reflects this change.

1.2 Benefits and Rewards

In my many years as Director of an actuarial work/study program, I have interviewed hundreds of students who have chosen to be actuaries. They all have one thing in common—*they all love mathematics.* Here is what some of them, and some of their employers, have given as reasons for their career choice.

Q Did you ever consider working in a non-actuarial field of applied mathematics (such as engineering) and if so, what tipped the scales in favor of an actuarial career?

Twenty-five percent of all respondents to the survey said "No." There was no doubt in their minds that all they ever wanted to be was an actuary. The rest had considered other careers. Here is what some of them had to say.

Answer I am currently working in a non-actuarial field where strong mathematical and financial skills are highly valuable. Elements that persuaded me to leave the actuarial field were salary and opportunity at the top management level.

Answer Yes. Communications and media. But I found that an actuarial career provides a more secure job, a great work environment, a good reputation, excellent job opportunities, and diversification of tasks, especially at the entry level.

Answer I considered studying engineering. I decided to follow an actuarial career instead because I didn't like some subjects in engineering (chemistry) and because the business part of an actuary's job interested me.

Answer I considered studying engineering. But I like the fact that being an actuary means that you need to acquire knowledge not only in applied mathematics (the primary reason why we're all in this field), but also finance, economics, taxes, politics, and all those things make an actuarial career so interesting.

Answer I did consider many other fields, including engineering and medicine.

Answer I was thinking about studying mathematical economics. Learning more about the actuarial profession and how challenging it is made me change my mind, and I never regretted it.

Answer I initially was seriously considering going into pure and applied mathematics and even engineering, until I stumbled upon actuarial science. It was the combination of the high-level applied mathematics and business skills required in this field that finally tipped the scales in favor of an actuarial career. The fact that actuarial science led to a much more rounded career appealed to me immensely and really made all the difference.

Answer Not really—I've been gunning for this since Grade 10. The workload of an engineering student at university steered me away from that, and I didn't want to be a computer programmer for my entire life.

Answer Yes. Statistics. But I felt a training in actuarial mathematics was broader and that it would be easier to switch from actuarial mathematics to statistics than the other way around.

Answer Yes. I applied to engineering. I then chose to become an actuary because it is more of a big-picture profession than engineering.

To be an actuary you need to have a long-term vision. You need to understand trends in the economy and be able to predict where the economy will be moving in the future. The concepts and theories you learn in statistics train you to think critically, to analyze, and to recognize patterns and trends.

Engineering is a more technical field and is not as conceptual as actuarial mathematics and statistics. I'm a big-picture man, and I believe that in the actuarial profession you get to see a lot more of the picture sooner. I assume that this training can also be applied to other fields in the future. It is a way of thinking and goes beyond technical knowledge.

Answer I haven't so far, but I'd like to keep my options open. The biggest stumbling block would be to realize how much effort I've put into the SOA exams to become qualified as an actuary and then ask myself, *Do I really want to ditch everything I've done for my career, put more time into studying something else, and take a 30% drop in salary?*

Answer I thought of being a teacher, but decided I didn't have the patience for that and I was drawn to a rotational-program setting at an insurance company so that I could have the exam support and variety of rotations. I would consider being an adjunct college professor or teaching an exam review class.

Answer Yes. But I decided to go in actuarial science because it was something less well-known to me and I found that to be a real challenge.

Answer I did consider it, but the job market favored actuaries at the time.

◆◆◆

1.3 A Typical Day

Let us take a look at a day in the life of an actuary. What are the typical tasks, and how does the day evolve? Obviously the answers depend on the nature of the company and the seniority of the actuary.

Here is what several actuaries and actuarial students had to say about this in the survey:

Q Describe a typical day in the life of an actuary.

Answer Reserve valuations. Asset and liability management. Dynamic capital adequacy testing. Pricing.

Answer Reading, replying and sending e-mail, letters and phone-mail. Keeping in touch with the daily activities of my clients and current economic developments.

Talking many times a day with the consultants I work with to keep track of the many projects going on and address issues if necessary. Producing reports of different kinds when a consultant has to meet with a client, depending on the client's needs and what the consultant wants to show them. Calculating performance figures from the different managers investing money for a client's fund, reviewing their historical performance and comparing it with a universe of funds and benchmarks. Following up on previous reports prepared for clients that need to be updated for the coming quarter. Verifying trust statements at the end of the month to make sure there are no discrepancies with the manager's data. Carrying out all kinds of calculations that are required by the consultants in their work with clients. Lots of teamwork.

Answer In the pension consulting industry, a typical day includes many phone calls with clients on subjects as varied as plan funding and investments, tax legislation, particular situation of given plan participants, union negotiations, benefit improvement, accounting treatment of pension plan, etc. Also, peer review of actuarial valuation results, planning and management of projects, business development, formal or informal training, internal or client meetings. It's rarely nine-to-five.

Answer A normal day in the life of an actuary at my level involves a lot of work with computers. Checking data, using programs to calculate liabilities for pension funds, personal calculations, all that can be done in a normal day. It is also not unusual to have training sessions on hot issues or new tools.

Answer I get to the office and check the e-mail and voice-mail messages. In the morning, I tend to work on projects until lunchtime and to contact my clients when problems arise. In the afternoon, I often have meetings with teams or clients, and I then keep on working on specific projects with different people.

Answer Consulting in group health insurance: technical work on actuarial valuation of post-retirement benefits. Core consulting: renewals, review of financial reports, benefits redesign, analysis of insurer's quotations on group insurance benefits. General advice to clients about current issues on group insurance benefits: phone calls, client meetings.

Answer For an actuarial intern, there is no such thing as a typical day. The tasks vary by intern and company but usually start with daily routine jobs such as updating data, checking the results of jobs run the previous day, and meeting with your supervisor. The remainder of the day is spent working on one or possibly several projects you've been assigned. Having junior status, an intern may work for more than one actuary and is often asked to run illustrations, compute premiums, search for data, make graphs, etc.

Answer There aren't too many typical days. Every day has some new wrinkle or challenge. Things that are done pretty much every day are working with spreadsheets to perform actuarial calculations, checking the reasonableness of the results of calculations (Are results reasonably consistent with your prior expectation of what the results should be?), communicating with both actuarial and non-actuarial coworkers in person, by phone, or by e-mail. And during exam season, studying for exams if you're still taking them.

Answer Here is an account of a typical day at the office. It's basically a 10-hour day:

8:00 Walk to the office.

8:30 Arrive at the office; read e-mail and news.

9:00 Finalize calculations for the report to client ABC; give directives to assistant.

10:30 Preparation for meeting with client A at 1 p.m.

12:00 Lunch with investment manager of firm.

13:00 Meeting with client A: presentation of the report submitted three days ago, discussion of the next steps and recommendations, and answer questions.

14:30 Prepare memo to client A following meeting concerning issues raised.

15:00 Debriefing with manager for client.

15:15 Consult voice-mail and e-mail.

15:30 Peer review report for client B.

16:30 Help junior analyst with calculation program for client C.

17:00 Contact Trust D for trust statement figures as of mm.dd.yyyy.

17:05 Search for client E: investment manager for an equity mandate.

17:55 Time entry for the day.

18:00 Go home (and study for actuarial exams!!!).

Answer Internship in a pension consulting firm: Every day is different. Different projects and obstacles to overcome. Challenging. It's hard to adjust between school and work routines. When beginning an internship, I often find myself very restless because I am not used to sitting in one place for long. At school, I never sit in one place for more than an hour.

Answer I arrive at the office at 7:30 a.m. I am usually the first one there, and I enjoy the quiet time to go through my e-mail, do some deep thinking, and plan the day's work. I am in the corporate actuarial department. We set valuation policy for the company or, more accurately, develop our company's interpretation of the valuation standards set by regulators and the Canadian Institute of Actuaries [CIA]. I am currently working on standards for applying the new *Consolidated Standards of Practice* to our valuation.

- *E-mail.* The first thing I do in the morning is to read my e-mail. I send an immediate response where I can, delete any notes where no further action is needed, store notes that form part of a discussion thread, and print anything that I need to spend more time on during the day.
- *Calendar.* Next, I check my calendar to see what meetings I have scheduled. Meetings can be a very significant portion of a working day, and if I have a memo or some other piece of work due that day, I need to do some short-term planning on how the work will get done on time. At this point I decide what I will actually do during the day. This will include meetings, project work, occasionally production work, and research. Project work is a catchall phrase for deliverables that take longer than a day. This could include developing standards for valuation, implementing a new computer valuation system, and collecting and coordinating data from different business units in support of a corporate decision. There always are one or two projects on the go that can absorb any available time in a working day not taken up by short-term requirements. Production work is usually tied to a particular time of the month or year, and relates to reporting requirements of one kind or another. My production work is to examine and analyze the source of earnings reporting for the company. Research means reading some of the CIA or OSFI [Canadian Office of Superintendent of Financial Services] papers that have been prepared for our education. Most of this is directly relevant to my current job since my department interprets these papers for the company.
- *The Rest.* The rest of the day is spent doing the work I have planned. My door is open, and the plans I have laid out are easily derailed if something comes up with a higher priority, such as a question from upper management.

Answer Consulting with clients of all sizes on a wide range of benefits-related issues including pension plan redesign, valuation, accounting, compensation, and expatriate benefits coordination.

Answer This greatly depends on the level of responsibility held by the actuary, the size of the organization in which the actuary works, and the type of company: life versus P/C [property and casualty], consultant versus insurer, and so on. The answer also depends on the period in question. For example, year-end will keep corporate actuaries very busy, while no overtime may be required the rest of the year. In any event, a day in the life of this actuary (meaning me) goes something like this. Bear in mind the following background information: I currently work for a small P/C reinsurance company (five employees), with both actuarial and underwriting responsibilities.

A typical day will see me coming in the office at 8:15 a.m. and leaving at 5:45 p.m.

Most of the day I work with Microsoft Excel. My work involves rating (calculating reinsurance premiums), production of reports, or corporate functions such as calculating IBNR [incurred but not reported] loss reserves, doing DCAT [dynamic capital adequacy testing] work, and analyzing quarterly financial information. Knowledge of Microsoft Word and Microsoft Access is also required, since we often write memos and reports and all our data is stored in Access.

Lunch is usually spent at my desk, reading e-mail, newspapers, or trade magazines, in order to stay abreast of current events in the world and in the insurance industry in general. From time to time, I may go out for a lunch meeting with a client or broker. Some travel is required from time to time.

Answer The day in the life of an actuary depends on a variety of circumstances: insurance versus consulting, life versus P/C, big company versus small, traditional role versus nontraditional role, and especially the line of business the actuary is involved with—and even that can vary from day-to-day! Actuaries I have met have handled pricing, reporting, risk management, reinsurance, and corporate and industry issues. Some are in non-actuarial roles like underwriting and senior leadership positions. Some work on group benefits (long-term disability, short-term disability, life, accidental death and dismemberment), some work on annuity products (fixed and variable), some work on life products (term, variable universal life insurance, universal life insurance), some work in investments, etc. I don't think that there is just one way to describe an actuary's day!

Answer In consulting: phone-mail, client calls with specific issues, tight deadlines, challenging work.

Answer In my experience, actuaries tend to while away their days solving problems. I believe a typical day for an actuary is made up of four basic functions.

- *Definition.* First, actuaries must carefully define the particular problem they are planning to tackle.
- *Research.* Actuaries must then research the problem. This research can range from using a library or the Internet to collect reference material to discussions with colleagues and coworkers.

- *Solution.* The third step involves the development, testing, and documentation of a proposed solution.
- *Implementation.* In the final step, actuaries seek approval and implement their solutions. Some problems are frequent but simple. In that case, the actuaries can complete all of these steps for a number of problems in a single day. Often though, actuaries have more complex problems that must be prioritized and addressed in a disciplined fashion. Dealing with a mixture of short- and long-term problems can also be seen as just another daily problem that actuaries can expect to have to deal with.

Answer I am currently involved in client support for an actuarial software package. I am also involved in the training of the users of this system. It is used for pricing, valuation, and other actuarial tasks. A typical day includes training of clients either in person, on the phone, or through the Internet. Our clients are located mainly in Canada, the United States, and Southeast Asia. I also answer e-mail from these clients regarding problems and questions they have with the system.

Answer I am an actuarial student working in the investment division of a company. I am the pricing actuary in my unit. I am responsible for pricing stable value products. I also work on product development. Once the market opens, I spend the first half of the day pricing cases. I am in close contact with the investment strategy group monitoring rate movements and analyzing certain risks. I am also in contact with sales, communicating rates throughout the day. My days are not planned because most of the cases are sent overnight, so every morning I go to work prepared for another challenging day.

Answer As a consultant in the asset consulting services group, my day is best described as working on different client projects, meeting investment managers to learn more about their team and investment process, delegating and supervising junior staff, meeting with clients to present reports—and studying during the evening for completion of the SOA examinations.

◆◆◆

1.4 Typical Projects

How do beginning actuaries spend their time at work, and how do these activities change as an actuary's career advances?

Q What are some of the typical actuarial projects on which you have worked, and what specific knowledge and skills were required? Please give some illustrative examples.

Answer   Union negotiations: they require strong analytical skills, a talent for multitasking, and the ability to work well under pressure.

Answer   Typical projects that I have been involved with include the production of reports, including writing, graphing, and editing charts; project management (requires good planning); communication with consultants (requires knowledge of the clients I work with, knowledge of Word, PowerPoint, and Excel); returns calculations (requires knowledge of the database, and basic financial mathematics).

Answer   Basic actuarial valuation: calculating the plan's liabilities from the data of the participants of the plan. Basic actuarial projects require rigor, methodology, and planning. Preparation of accounting disclosure and calculation of pension expenses: knowledge of accounting rules and their application.

Answer   I've worked on annual statements. A good knowledge of Microsoft Excel and pension plans was required. Being methodical and having good organizational and language skills are important. I've also worked on actuarial evaluations. The same skills and knowledge as for the statements were needed, plus a good knowledge of valuation software, as well as familiarity with the law and the valuation process (gain and loss, reconciliation, etc.).

Answer   Typical projects that I have been involved with included:

- *Valuations.* Actuarial valuations: determination of the present value of annuity benefits taking into consideration demographic factors (mortality, termination, retirement, etc.).
- *Reports.* Financial reports: understanding how balance sheets work, statistical knowledge, analytical skills, credibility notions, software skills (Fortran, Microsoft Excel).
- *Computing.* Software skills are crucial in the actuarial field. A good grasp of Excel, AXIS, Microsoft Visual Basic for Applications, and even APL are a great advantage as they are widely used in the field. The main project I worked on consisted of reviewing and updating a computation made in the valuation system of an insurance company. My work was very specific and involved many calculations, running illustrations, and analyzing results.

- *Products.* I also needed to have a good knowledge of the various products sold and their specific details. For example, if my results seemed irregular, my first instinct was to look up the product I was examining for distinct features such as product design or recent repricing.

Answer Typical projects that I have been involved with included:

- *Reserves.* Calculating reserves: needed knowledge of actuarial mathematics (life contingencies, theory of interest) and general structure of reserves, as well as computer software. Knowledge of professional standards of practice is also needed.
- *Balances.* Calculating fund balances for retirement and investment products: actuarial knowledge of the theory of interest and computer software were essential. Knowledge of legislation regulating such products is also needed.
- *Design.* Design of insurance and investment products: knowledge of the different mechanisms of insurance products, knowledge of different investment products, rules and regulations regarding those products, computer software, and communications skills when working with others were essential.

Answer Typical projects that I have been involved with include actuarial valuations of pension plan liabilities; costing of plan benefit changes; and pension expenses. The skills required for these projects were basic technical skills: mathematical, actuarial and accounting rules, knowledge of internal valuation software, and knowledge of laws affecting pension plans. I have also written reports to clients: letters, actuarial valuation reports, investment manager monitoring reports, etc. The skills required for this type of work are the ability to translate complex issues into understandable words, writing skills, and communication skills.

Answer Reserve valuations, year-end and quarterly pricing, new products, modification of current products, DCAT [dynamic capital adequacy testing], business projections for the next five years, performed once a year, and MCCSR [minimum continuing capital and surplus requirements] calculations.

Answer Installation of new valuation systems, project management, ability to reconcile old and new valuation systems by results (actuarial, analytical abilities), ability to influence an area over which you do not have direct control, sense for when an approximation is OK, pricing of retirement savings product, knowledge of corporate pricing targets and practices, software skills, ability to seek and accept input from producers, ability to reconcile conflicting priorities of sales, ability to deal with management and the corporate office, and the ability to build consensus.

Answer Most are in line with post-retirement benefit valuations. Specific knowledge required: applying discount and mortality data to benefits scheduled for a future date.

Answer Typical projects that I have been involved with included:

- *Valuations.* Pension plan valuations. They are needed to ensure that the retirement benefits promised to employees by their employers are available for their retirement life. A valuation calculates the value of those retirement benefit promises (pension liabilities) and compares them to the assets invested. A fully funded pension plan is a plan that currently has a level of assets sufficient to cover its pension liabilities. Skills required: actuarial background to calculate the required values; programming skills to understand/program/run the system on which the liabilities are calculated; analytical skills to check, compare, and compile results; up-to-date knowledge on current market and economic issues used to set and understand the assumptions used in the valuation.
- *Benefits.* Administering the benefits of expatriates working in various countries. Expatriates add another layer of complexity in benefits valuation since coordination is required between the host and home countries, as well as potential social security benefits earned in various countries.

Answer Typical projects that I have been involved with included:

- *DCAT.* A lot of work has been done recently on DCAT [dynamic capital adequacy testing]. Essentially, this is a financial model that projects the future financial condition of a company. The model can be deterministic or stochastic in nature. In my last three jobs, I have been involved to various degrees with this. This type of project requires good understanding of accounting concepts (projection of balance sheet and investment income), investment concepts (calculation of market and book value of investments under various economic scenarios), financial concepts (calculation of corporate income tax), and statistical concepts (calculation of various probability scenarios). Developing appropriate business knowledge through finance, economics, investment, and management courses can never be stressed enough.

- *Computing.* Other projects that I have been involved with usually only require a good understanding of actuarial concepts, acquired through coaching and through the examination process. Expertise with Excel is always a must. So are other computing skills: Microsoft Visual Basic and SAS being the most common ones.

Answer Typical projects that I have been involved with included:

- *Annuities.* I've worked on developing new annuity products and riders (i.e., product management: seeing an idea develop into a real product that is sold to contract holders). Within that process, I have worked with all business areas (compliance, legal, marketing, systems, etc.) to get an idea into a working product.
- *Ratemaking.* Other projects included setting the credited rates for our various fixed and variable annuity products.
- *Profitability.* I have worked with in-house actuarial software to examine profitability.
- *Verifications.* I have verified client illustrations to verify that what is being shown to a client for an annuity product's sub-account growth and death benefit calculations is accurate.
- *Reviews.* I have also done product reviews of our existing products to validate the pricing.
- *Economic Value.* I've worked on economic value—determining which areas of the company are contributing what value to our theoretical stock price.
- *Reinsurance.* I also worked in reinsurance where I dealt with reinsurance intermediaries and brokers to renew contracts. This also involved assessing the risk within our existing contracts.

Answer Renewal analysis (group insurance); financial statement analysis (group insurance); reserves analysis; post-retirement benefit valuation; report writing; various types of research; preparation of benefit statements; policies and booklets verification. Knowledge and skills: computer knowledge (programming, Microsoft Word and Excel), communication skills (in French and English), writing skills, planning ability.

Answer Actuarial valuations (knowledge: methods for valuing liabilities); accounting procedures (knowledge: basic accounting); calculations (knowledge: laws and regulations, ability to draft reports, good reading comprehension); plan design (knowledge: industry trends).

Answer Typical projects that I have been involved with included:

- *Reserving.* Standard reserving projects. Involved applying various development techniques (mostly triangular methods) to estimate ultimate losses, determining liabilities on unearned premiums, discounting loss payments, and calculating provisions for adverse deviation. Skills required: analytical skills, technical knowledge of actuarial methods, common sense, familiarity with types of insurance, lines of business, and coverages analyzed. An example might be the projection of asbestos and environmental liabilities arising from old, expired policies issued to commercial clients. Skills required: strong analytical skills, problem solving, creativity, curve fitting, stochastic modeling, computer programming, knowledge of legal environment.
- *Pricing.* Determined indicated overall premium change, calculated required change in base rates and relativities for various rating variables. Skills required: analytical skills, technical knowledge and understanding of actuarial ratemaking techniques, curve fitting, good judgment.
- *Benefits.* Special studies: impact of change in statutory benefits provided under accident benefits coverage for auto insurance. Skills required: strong analytical skills, problem solving, creativity, resourcefulness.

Answer An experience study is a typical but simple actuarial project. An actuary may be required to analyze an experience as often as each calendar month. To complete this type of project on such a frequent basis, the actuary generally keeps the process simple and may rely heavily on computer systems. This requires the actuary to know about probability and statistics as well as mortality table construction, finite mathematics, survival models, and computers. The calculation of an embedded value for a company or block of policies would be a complex problem and could take a dedicated team of actuaries a number of months. The actuary would have to know about the relevant methods for dealing with asset and liability data, selecting assumptions and reserving methods to apply to this data and implementing computer systems to translate all this into a simple range of values.

Answer Typical projects that I have been involved with included:

- *Illustration.* Programming of an illustration system. Skills required were programming, analysis, and client contact. Training of clients using the company's system. Skills required were knowledge of the system and training capabilities.

- *Reports.* Preparation of actuarial reports for court cases. Skills required were knowledge of laws and capability to write the reports.

Answer Cash flow testing, economic value benchmarking, product development and pricing. Helpful courses: life contingencies, theory of interest, knowledge of fixed income securities.

Answer An actuarial background is not a prerequisite to work in the asset consulting services group. Some of my colleagues have a finance background. In fact, the projects I work on are not purely actuarial projects.

Answer Typical projects that I have been involved with included the following:

- *Assets.* How should the assets of a pension plan be invested? These projects are mostly worked on by actuarial people. They require a knowledge of both the liability and assets sides of a pension plan: demographics, financial results, investment markets, etc.
- *Investment.* Review the pension plan (statement of investment policy and procedures).
- *Management.* How to implement an investment policy, how many investment managers to assign to each asset class, what kind of investment managers to select: large/midsize/small capitalization, value/growth/core investment style.
- *Personnel.* Selection of investment managers for each asset class (Canadian equities, U.S. equities, international or foreign equities, fixed income, etc.). A management structure and the manager selection require a good knowledge of the institutional investment market; you need to know the players, their investment process and style, and so on.
- *Monitoring.* Monitor the investment performance of each manager. It requires a good knowledge of their style as well as how the markets are performing in order to really understand their numbers and be able to explain their performance to clients. You need to know the team players in order to monitor any changes and turnover of people.
- *Mandates.* Put in place the appropriate paper documents between the pension plan and the investment managers.
- *Records.* Defined contribution record keeper selection.
- *Options.* Defined contribution investment options selection.

◆◆◆

Entry-Level Jobs

Q What are the responsibilities of new employees in actuarial entry positions in your company, and what are their typical tasks and salary ranges?

Answer Preparation of reports, letters, documentations, filing for clients. Technical knowledge to accomplish the work. Communicating effectively with consultants by phone and mail. Quality of the work done (as accurate as possible). Passing exams. Salaries range from $35,000 [2002] to $50,000 [2002] CAD [Canadian dollar].

Answer Entry-level employees will normally work on plan participant data that will be used for actuarial valuation purposes. Preparation of annual statements. Preparation of various worksheets, projection of calculations, etc.

Answer With three actuarial exams, the starting salaries are about $45,000 [2002] CAD. At the starting level, actuaries are more technical experts. They are in charge of the computer work and getting to know their clients' plans, laws, etc.

Answer Technical analysis, renewal, financial reports, database statistics on current topics. Salaries are between $30,000 [2002] and $35,000 [2002] USD [United States dollar].

Answer New employees in a company are usually actuarial assistants or analysts, and must focus on learning all they can about the operations of the company. Knowing and understanding how and what the company does are crucial. Responsibilities consist of testing products, running various scenarios, researching and observing trends in the market and how the company fares, gathering data, computing premiums, etc. As experience is gained, more responsibilities are given. Salaries will vary according to the number of professional exams, but should rank between $30,000 [2002] and $35,000 [2002] USD.

Answer New employees tend to work on specific projects or may be assigned tasks that are periodic in nature. Difficult to describe or list specific tasks since they are usually company-, department-, or manager-specific. New students are expected to be able to develop their skill at judging reasonableness of results, using the experience they gained by working with their managers. Students may be asked to summarize results from reserve calculations to see if they were done correctly. They may also be asked to participate in research studies in which they may perform many data-massaging exercises. Salaries for a beginning student in the United States seem to be between $45,000 [2002] and $50,000 [2002] USD.

Answer *Responsibilities:* Limited responsibilities. New employees are expected to:

- Understand what is asked of them (by asking questions, taking notes, etc.) and return what is expected (quality job, list of questions that arose while doing a job, etc.).
- Acquire basic technical skills, adapt to and learn internally used software, procedures, etc.
- Participate in various portions of projects, supported by more senior employees.
- *Typical tasks:* Compile, clean, and analyze data. Programming required for actuarial valuations. Help in preparing reports (stats, other calculations, etc.).
- Salaries vary by province, city, and country. In Montreal, salaries would be between $30,000 [2002] and $35,000 [2002] CAD, depending on the number of actuarial exams, company size, etc.

Answer Responsibilities involve number-crunching. By that I mean doing all basic calculations involved in a project. From calculating a projected cost to a simple projection of a cost. In my case, with a bit more than a year of experience, I often need to value future benefits. This involves calculating the present value of future benefits for all employees of a given client.

Answer Entry-level positions are usually actuarial analyst positions. Candidates are expected to be actively completing SOA exams and usually have passed the first few courses when hired. Responsibilities: working closely with senior analysts and junior consultants on a wide range of client projects. Project responsibilities usually include analyzing data and programming, calculating pension plan liabilities, compiling and analyzing actuarial valuation results, answering various day-to-day client questions relating to pension plans, special projects. I'm not sure what starting salaries would be. In Toronto, probably somewhere around $55,000 [2002] CAD, depending on the number of exams passed.

Answer My company does not have entry-level positions. However, my previous company (a P/C insurance company) does have such positions. Typically, a starting salary will depend on the number of exams passed. University graduates with two exams could be making about $45,000 [2002] CAD in Toronto. Salary ranges vary considerably depending on which city you live in. Responsibilities would include the production of routine reports, the preparation of ratemaking or reserving, Microsoft Excel spreadsheets for analysis by more senior actuaries (and eventually some analysis with the help of the senior actuary), programming, and data entry. The quality of individuals will most often dictate how quickly their salaries rise, and how quickly more responsibilities are assigned to them.

Answer Entry-level students can play a role throughout the company in a variety of positions. Typical starting salaries would probably be about $50,000 [2002] CAD, assuming the individual had passed one exam. Students can work in financial reporting roles doing the monthly, quarterly, and annual output, and in product development roles assisting more experienced actuaries. In our company, an entry-level student might be placed wherever assistance is needed. Entry-level students are expected to build upon their technical and communication skills as well as pass exams.

Answer *Responsibilities and tasks.* gradually communicate with clients, write reports and letters, carry out calculations and do research, prepare internal presentations, and write articles on "hot" subjects.
Basic valuation work. individual calculations.
Responsibilities. analyze and price or reserve casualty, life, or health insurance products. Salaries range from $42,000 [2002] (base salary for candidate with no prior internship, no actuarial exam, weak grade point average) to $54,000 [2002] CAD (candidate with advanced degree, prior internship, strong GPA, and three actuarial exams: $3,000 [2002] CAD per exam).
Programming. In Microsoft Visual Basic.

◆◆◆

Intermediate-Level Jobs

Q **What are the responsibilities of employees in intermediate actuarial positions in your company, and what are their typical tasks and salary ranges?**

Answer At this level, an actuarial employee will be responsible for interfacing with clients on a daily basis, as well as peer review, the management of projects, and the supervision of junior staff.

Answer More consulting: meetings with clients, providing advice, preparing of documents for presentation to clients, reviewing of more junior technical work. Salaries range between $45,000 [2002] and $50,000 [2002] USD.

Answer Intermediate actuarial employees are usually assigned specific projects requiring good problem-solving skills. For example, they could be asked to come up with a new way of computing certain data that are presently time and cost consuming, or may be asked to design a new approach for calculating reserves for a new product with nontraditional features. Salaries may range from $35,000 [2002] to $40,000 [2002] USD.

Answer   Intermediate actuarial employees should not only be able to handle routine tasks and mathematical model building that an entry-level employee would need to do, but should also be able to significantly modify or redesign models. They are expected to have a more complete understanding of the industry practices and regulations, and be able to use their judgment in applying these standards to working situations. Decision-making ability should be more developed. Salaries may range from $60,000 [2002] to $75,000 [2002] CAD. Similar ranges in USD apply to students in the United States. New FSAs [Fellows of the Society of Actuaries] usually start at about $75,000 [2002] to $80,000 [2002] CAD.

Answer   Intermediate actuarial employees have broader responsibilities. They are expected to be able to lead small to medium-sized projects through all the steps and train junior employees. Typical tasks include the preparation of reports, and peer review of calculations and programs. Salary ranges vary by province, city, and country. In Montreal, salaries range from $40,000 [2002] to $50,000 [2002] CAD, depending on the number of actuarial exams, company size, etc.

Answer   Assistant actuaries (first level after becoming an FSA) should be able to write technical memos presenting assumptions, data, discussions, and conclusions for an audience of actuaries. For example, they should be able to write a note on how the investment assumption for a business unit was developed, describing the asset classes, the starting yield curve, the development of PfADs [provisions for adverse deviations], the reinvestment assumptions, and any planned changes to the investment policy. They could also be valuation managers (persons who actually run the valuation programs and determine the reserves, under the management of a valuation actuary).

Answer   Intermediate actuarial positions in my view would be for analysts with three to five years of experience, near qualification as FSAs, and who are usually ASAs. Salaries in Toronto probably are between $65,000 [2002] and $80,000 [2002] CAD, depending on the number of exams and performance. Responsibilities include working closely with junior analysts and consultants on a wide range of client projects: coordinating and managing projects, delegating project tasks to junior analysts, and preparing final projects for presentation to clients. They are involved in day-to-day client queries and projects, and participate in relevant client meetings and discussions.

Answer My current company does not employ intermediate actuaries. In my previous company, this type of position was held by people with five to
seven P/C [Casualty Actuarial Society] exams (out of nine), and/or four
to five years of experience. Salaries depend on years of experience and
successful exams, and range from $40,000 [2002] to $50,000 [2002]
USD, or more. Intermediate actuaries tend to be responsible for specific
actuarial projects (rate review for a given product and province, review
of the IBNR [incurred but not reported] loss reserves, planning premiums and loss ratios for the budget, monthly review of results), subject
to the supervision of a manager (typically, a Director or Vice-President).
They may or may not have the help of a junior actuary. They also tend to
be the experts assigned to company-wide projects, whether the project is
an IT [information technology] (new rating engine) or business project
(review of claims reserving practice by claims department).

Answer Most intermediate Students [actuarial employees who have passed three
to six exams] probably make around $60,000 [2002] to $80,000 [2002]
CAD. They work in any of the various areas of the company and are
expected to be further honing their technical and, especially, communication skills. Typical tasks include a more advanced role in various
areas of the company since they are expected to understand the corporate structure and have at least one to two years of experience behind
them. Often they are encouraged to be managers of summer interns to
get some management experience.

Answer Typical activities of intermediate actuaries involve communication with
clients, preparation of reports, verification of calculations, sales, presentation to clients, training of other employees, and billing.

Answer Their activities include the delegating of work to junior staff, checking
their work, and dealing with clients.

Answer Same as for entry-level positions. Salaries range from $50,000 [2002] to
$125,000 [2002] USD.

◆◆◆

Progression of Responsibilities

Q What are typical SOA and CAS career paths and where should successful actuaries or actuarial students be at age 20, 25, 30, 35, 40, 45 in (a)
SOA, (b) CAS?

Answer At 20: at school. At 25: junior consultant. At 30: pre-senior consultant. At 35: senior consultant. At 40: advanced senior consultant. At 45:
responsible for major clients, line of business, direction, etc.

Answer I don't think there is such a thing as a typical career path. For SOA, actuaries should have a firm client relationship with clients by the time they're 30. At 45, they should be established as client managers, responsible for high-level work and the relationship with clients.

Answer At 25: SOA graduate with the first three or four exams. Then continue to write exams while working and finish before 30. At that age, actuaries should be familiar with the technical concepts and begin to be relatively autonomous in establishing what needs to be done on different projects. At 30, they should be able to review the work of junior students and have their own clients. At 35, they should be senior consultants.

Answer At 20: finishing an undergraduate degree in statistics or actuarial sciences and have at least written the [equivalent of the current pre-module examinations] exam. At 25: be at about Courses 4 [2002] or 5 [2002], and have spent one or two years as an actuarial assistant. At 30: have completed all courses and have gathered five to eight years of experience in one or more companies and hold an Assistant Manager's position. At 35: manager or director. At 40: permanent senior position, secure and confident in the position they are holding.

Answer In my opinion, this should be stated in terms of duration from when the first exam is attempted, rather than by age. People get into the field at different ages and different places; they have different average ages upon graduation from college. Thus, it is not uncommon for someone to get their FSA prior to age 25 in the United States, whereas it is less common in Ontario because Ontario students graduate from university when they are between 23 and 24, instead of 21 or 22. I have met people who didn't start taking exams until their 30s because they switched careers. Most students should get their Fellowship about 8 to 10 years after they started taking exams. The average age of new FSAs is usually in the mid-30s, although the SOA wants to reduce the exam travel time and, indirectly, the average age of new FSAs. They want to do this, but I doubt it will happen.

Answer I will cover only the SOA exams. At 20, start the exams if you want to be an actuary. At 25, you should have completed Courses 1 [2002] through 4 [2002]. During the first few years of your actuarial career, you will be a junior actuarial analyst. At 30, you should at least be an ASA. You will be a senior actuarial analyst or junior consultant. At 35, you should be an FSA or have decided whether you want to continue writing exams. Life consists of more than SOA exams!!! You should be an intermediate consultant. At 40, you should be a senior consultant.

Answer I would like to have finished all of my SOA exams by the age of 24. After a three-year university program, a solid goal is to have passed [the equivalent of] four [pre-module] exams.

Answer At 20, you should be in university and have started the first two or three exams. At 25, you should ideally have finished your exams and should be waiting for the completion of your PD [professional development] requirement credits. At 30, you should have two or three persons to whom you delegate work and start helping them build their knowledge. At 35, you should focus on networking and meeting people, start bringing clients to your consulting firm and maintain relationships with existing clients. At 40, you should probably be at the peak of your responsibilities.

Answer At 20: actuarial student. At 25: senior actuarial student. At 30: FSA. At 35: Associate Actuary. At 40: Assistant Vice-President. At 45: Assistant Vice-President or Vice-President.

Answer From the SOA point of view: At 20: in university. At 25: starting out, passing exams, gaining experience at an actuarial firm, deciding on insurance versus consulting, SOA versus CAS, is or is close to being an ASA. At 30: is or is close to being an FSA, settling into the actuarial field with preference for insurance or consulting, SOA or CAS. At 35: twelve or more years of experience. Consultant level with expertise in a preferred field. Providing valuable advice to clients on a wide range of client issues, and a good source of intellectual capital for peers.

Answer I will answer this question from the CAS perspective.

- *Student.* Typically, at 20, you will still be in university. You will hopefully get some summer work experience, if not working for an insurance company, at least getting some exposure to the office world. You should be planning to write a few actuarial exams while in university to show prospective employers your willingness to write exams, and your capacity for writing them successfully.
- *Intermediate.* At 25, you should be making the transition from entry-level to intermediate actuary. You should have written several exams by now, including basic ratemaking and reserving (although not necessarily passing them), which will prove invaluable in the new responsibilities being handed to you.
- *Associate.* At 30, you should be an Associate, even a Fellow if you are one of the more gifted. This is the point in your actuarial career where you are handed management responsibilities. Although everyone wants to be a manager, very few understand what is involved. If working for a

good company, the actuary will have been sent to some form of management and other business-training seminar. But the very motivated individuals will not rely on the company, and will read up on these subjects at home.

- *Vice-President.* At 35, most CAS Fellows are Vice-President or its equivalent such as partner in a consulting firm (at least, in Canada). Responsibilities start shifting from the pure actuarial areas to the areas of company management and client management.
- *Career Peak.* At 40 and 45, your level of responsibilities will slowly increase, but essentially, things will remain the same until retirement.

Answer I cannot respond with respect to the CAS, so the answers below are with respect to the SOA.

- *College.* At 20: Taking college courses towards a mathematics degree or actuarial degree. Investigating internship opportunities. Planning to take one or two exams before graduation.
- *Work.* At 25: Working at a company with one to three years of experience. Have passed two or more exams.
- *Almost FSA.* At 30: Working for a company with 5 to 8 years' experience. Be close to attaining FSA if not already an FSA.
- *FSA.* At 35: Have the FSA designation, have 10 to 13 years of experience, have a staff working for you, have a more prominent position and be out of the "rotational" student program. Know your specific area of interest or track the one you want to pursue in depth.
- *Leadership.* At 40 to 45: Have a significant leadership role within the company and a staff working with you. Be accessible to newer students in the program who want advice.

Answer In my answer, I will focus on a CAS career. At 20: nice to have passed at least one exam. At 25: two years of experience and at least four exams. At 27: five exams. Very good candidates will have access to a managerial position. At 30: will probably be a Fellow by this age—if not, no problem, but focus on finishing the exams. At 30 to 35: outstanding candidates will have access to senior management positions. Over the first ten to fifteen years of an actuary's career, it is not uncommon for a person to have worked for several employers.

Answer I will describe a typical SOA path. At 20: junior staff in consulting firms or insurance companies. At 25: almost a consultant. At 30 and above: senior consultant for clients and relationship manager.

Answer   Here is a typical CAS career path:

- *School.* At 20, students are still in school, completing their Bachelor's degree (or Master's degree, even though it is not required in the actuarial field). While in school, students generally start taking actuarial exams. A successful student should have passed the first two exams before graduation and should have had at least two internships related to the actuarial profession or insurance industry.
- *Analyst.* At 25, students should have a year or two of experience and be well established as an actuarial analyst. At that point, a successful student would have passed 5 or 6 actuarial exams.
- *Almost Fellow.* At 30, actuaries have typically been exposed to various aspects of the actuarial profession and have expanded their experience to pricing and reserving different lines of insurance. They also have analysts reporting to them and should be close to obtaining their Fellowship (if not done already).
- *Vice-President.* By 35, actuaries should definitely have their Fellowship and be in a management position (either as a senior consultant in a consulting firm or a Vice-President, or Assistant Vice-President, in an insurance company).
- *Partner.* At 40, a successful actuary would be a partner in a consulting firm or an officer in an insurance company.
- *Retired.* At 45, a *very* successful actuary would retire.

Answer   At 20: in university. At 25: actuarial student, 35 hours per week on the job and 40 hours per week studying. At 30: new Fellow, supervisor or manager of a few actuarial students and clerks, or a highly technical position without direct reports. At 35 and above: continually increasing responsibility, demonstrated by increased staff and budget or required technical knowledge.

Answer   I am not aware of requirements for CAS. At 20: in university writing exams. At 25: out of university in a junior-level position. At 30: almost done with the exams and with some supervisory responsibilities, changing departments on a biannual basis. At 35: done with the exams and with more supervisory responsibilities. At 40 and 45: same as 30 and 35, but settled into a department.

Reality Check

As you read this book and reflect on the answers to many of the questions answered by the respondents, you might find it interesting to compare how your own projective plans match up with those of many of the highly successful actuaries whose views and choices persuaded me to update this book. Try to imagine which of the following accreditation paths might correspond to your career expectations. Keep in mind, though, that the speed of accreditation and the number of Fellowships obtained are only partial measures of professional success. Note that the listed sample profiles are not necessarily of actuaries who have contributed to the Q&A sections of this book.

- All actuaries completed a Bachelor's degree in mathematics and/or actuarial science.
- Only one actuary went on to Graduate School and completed an MBA [Master of Business Administration].
- Three actuaries also obtained CFA [Chartered Financial Analyst] diplomas.
- All actuaries who chose careers in insurance became SOA 2009-2010 Compliant, discussed on the SOA website, and so did one actuary working in property and casualty insurance. (See the website www.soa.org/files/pdf/2010-florida-griffen-57.pdf to learn what "compliance" is all about.)
- One actuary became a Member of the American Academy of Actuaries.
- Three actuaries chose careers in banking and investment.
- Two actuaries left the actuarial profession.

The following list shows the varying accreditation paths of some of the fifty respondents.

1. Graduation (1990); Accreditation: FSA (1993), FCIA (1993); Specialization: Insurance
2. Graduation (1993); Accreditation: ASA (1992), FCIA (2000), FCAS (2001), Compliant (2009); Specialization: Consulting
3. Graduation (1996); Accreditation: FCIA (2000), FCAS (2000); Specialization: Insurance
4. Graduation (1997); Accreditation: FSA (2002), FCIA (2002), Compliant (2009); Specialization: Consulting
5. Graduation (1997); Accreditation: FSA (1999), FCIA (1999); Specialization: Insurance
6. Graduation (1998); Accreditation: FSA (2004), Compliant (2009); Specialization: Consulting
7. Graduation (1998); Accreditation: ASA (2002), Compliant (2009); Specialization: Consulting
8. Graduation (2001); Accreditation: ASA (2002), FSA (2002), FCIA (2003), Compliant (2009); Specialization: Insurance

9. Graduation (2001); Accreditation: FCIA (2010), FCAS (2010); Specialization: Insurance

10. Graduation (2002); Accreditation: FSA (2006), FCIA (2006), Compliant (2009); Specialization: Consulting

11. Graduation (2002); Accreditation: FSA (2007), Compliant (2009); Specialization: Consulting

12. Graduation (2002); Accreditation: FSA (2004), FCIA (2004), Compliant (2009); Specialization: Consulting

13. Graduation (2003); Accreditation: FCIA (2007), FSA (2007), FIA (2010); Specialization: Consulting

14. Graduation (2004); Accreditation: FCIA (2007), FCAS (2007); Specialization: Consulting

15. Graduation (2004); Accreditation: FCIA (2008), FCAS (2009); Specialization: Insurance

16. Graduation (2005); Accreditation: FSA (2008), Compliant (2009); Specialization: Consulting

♦♦♦

1.5 | Mathematical Skills

Here is what the respondents to the survey had to say about the basic mathematical knowledge they require in their daily work. They also commented on the connection between theory and practice. What links are there between the actuarial examinations and their required working knowledge of mathematics, finance, economics, and other special subjects such as risk theory, loss modeling, and stochastic methods?

Q **What general mathematical competencies are required by an actuary? Give some examples and relate them to the SOA or CAS examinations.**

Answer Return on asset calculation (Course 2 [2002]), retirement plan methodology and characteristics (Course 5 [2002]), statistics related to risk (Courses 1 [2002] and 3 [2002]), pretty much all of Course 6 [2002] for me (as a junior) in asset management, basic financial mathematics (Course 2 [2002]).

Answer Problem-solving, but this has nothing to do with any university course or actuarial exam.

Answer Calculus is needed for the first actuarial exam. Financial mathematics is very useful in the day-to-day work as well as for the exams (tested on more than one exam). The whole of actuarial theory is based on statistics, so it is, of course, a required competency.

Answer For the first exam, you need a lot of basic probability and calculus competency. The second exam is more about financial mathematics, macro- and microeconomics, and finance. The general mathematical competencies required for this exam are mostly integrals and derivatives. After that, you will always be using a variety of mathematical competencies (again basic probability, integrals, and derivatives), but they will become more specific.

Answer Course 1 [2002] deals with basic probability and calculus competencies. An ability to deal with them and apply them to actuarial problems is crucial. In general, a deep understanding and competency in probability and statistics is essential to passing SOA and CAS examinations since they are the foundation of actuarial mathematics. A strong background in statistics is necessary.

Answer Knowledge of probabilities and statistical distributions, life contingencies, theory of interest, calculus, geometric series. The calculus is often tested via continuous distribution functions where integration of a function is required (Course 1 [2002]). Probabilities of people living and dying are combined with geometric series to create the mathematics of insurance and annuities (Course 3 [2002]). The theory of interest is used for the principles of interest discounting and accumulation (Course 2 [2002]).

Answer Theory of interest, life contingencies.

Answer Theory of interest is a must (time value of money). Probabilities are also very important.

Answer Well, everything that's mathematical in the exam syllabus. Plain and simple!

Answer Competency in calculus, statistics, algebra, probability is essential, especially for the early exams.

Answer Actuarial mathematics such as life expectancies, survival models and projections, annuity factors, regression analysis (e.g., calculating trends, building models, etc.). Calculus: background used in most programs and models. Statistics: always needed to calculate averages, medians, quartiles.

Answer This is a difficult question. The answer also depends on the level of sophistication reached in the various companies. P/C companies in

Canada are small and not a lot of complex mathematical models are built. I know of one or two companies working on that front, and they have hired a person with a Master's degree in statistics to do the work. However, these people are supervised by actuaries. Advanced knowledge of calculus, statistics, theory of interest, life contingencies, and loss distributions are generally required to pass the first four exams. Past that point, at least on the P/C side, mathematical competency almost boils down to being able to add and multiply. Basic knowledge of the above is all that is required. And as I said, I find the same is true for our day-to-day life at work.

Answer CAS: Only basic mathematical competencies are required. Regression and modeling may be beneficial, but are not a must. It is a common mistake to believe that extensive knowledge of mathematics is required to be an actuary. However, one must like to work with numbers to enjoy being an actuary.

Answer Basic mathematical skills needed. The examinations helpful for a career are the ones that discuss the different methods for valuing liabilities, accounting, and finance.

Answer Calculus, probability, and statistics.

♦♦♦

Q Why do actuaries need calculus? Please give examples and relate them to the SOA and CAS examinations.

Answer Rarely used so far in my career, and if I happen to need a concept from calculus, I can easily find someone in the office who will be sharper than me on that subject. The more I advance, the less I see a calculus background as being useful at work. But I can understand that it is a great mathematics background to have as an actuary.

Answer They don't need it for most of their day-to-day work.

Answer Actuaries study things that change as part of their daily work. Calculus is the mathematical construct that is used to quantify, measure, and discuss how things change. I don't think anybody who truly understands change should have problems with calculus. People who have difficulty with calculus will probably lack the problem solving skills that are required of an actuary.

Answer Mostly for the exams. So far, I've never used it in my job.

Answer I don't believe calculus is actually used directly in the everyday life of an actuary, but it is a mathematical concept that needs to be understood by anyone who is said to be an expert in mathematics.

Answer They don't.

Answer From my point of view, calculus is only helpful for the first actuarial examination. After that, you will only use simple applications.

Answer Calculus is a basic tool used in probability that must be mastered. Questions arising in Course 1 [2002], for example, will deal with those competencies. Also, Course 1 [2002] will specifically ask calculus questions. Calculus is also a basic tool used in actuarial mathematics. In Course 3 [2002], for example, it is crucial to have a good grasp of calculus to successfully pass this course.

Answer Probability of paying a death benefit on any day required integration over a continuous distribution function—which is calculus. Also, trend analysis uses predicted rates of change, which is calculus. This mostly crops up on examinations in Courses 1 [2002] through 4 [2002].

Answer Knowing how to integrate or estimation using sums is the basis of most actuarial valuation formulas. Integrating is also the basis of modeling (e.g. using the normal or lognormal distributions).

Answer I do not find any direct application, although it could have helped in providing me tools for analysis and a workout for my brain.

Answer I love calculus; I always have. I think that knowledge of the properties of basic functions, continuity, and multidimensional spaces is essential food for thought. I think that the practical applications are limited. Basic calculus is tested in the Course 1 [2002] exam. Continuous life insurance premiums, coverage, and annuities are dealt with in life contingencies, Course 3 [2002], but I doubt that any of this is used in practice. It is still good conceptual training.

Answer Concepts like "rate of change when delta t is minimum." Things such as force of mortality. You need a calculus background to grasp exactly what it means. Personally, I think probability and statistics is much more important.

Answer Mostly for the exams. So far, I've never used it in my job.

Answer To evaluate continuous probability density functions, to evaluate continuous mortality functions.

Answer   Needed in order to understand the underlying models and processes. Although most actuaries don't sit around to derive and integrate all day, calculus is required to understand the underlying actuarial formulas, calculations, processes, etc. Computers do most of the work, but calculus is a basic building block. The syllabus on the mathematical SOA exams is always more technical than the skills you'll ever need in real life.

Answer   I'm not sure there is a great need for calculus in our day-to-day job. However, calculus will help form a problem solving mind set, I find. I have used calculus, personally, to calculate, for example, the average earning period for our unearned premium (simple matter of integral). This is a really basic Course 1 [2002] question.

Answer   They do not in their day-to-day work. However, taking calculus is part of having a general knowledge about mathematics. I would not discontinue calculus courses—or any other mathematics subject for that matter— because they are not used in our actuarial day-to-day work. As far as the exams are concerned, knowledge about calculus is needed to be able to answer the questions. That's it.

Answer   Personally, no.

Answer   Understanding the general formulas.

♦♦♦

In addition to calculus, Course 1 [2002] covered basic probability and statistics. Here is what working actuaries and actuarial students had to say about their view of the importance these topics:

Q   **Why do actuaries need probability and statistics? Please give examples and relate them to the SOA or CAS examinations.**

Answer   Laws of probability and statistics are useful in my work, but only basic concepts are needed on a daily basis. These concepts must be very well understood. For the rest, I consult books when necessary.

Answer   To understand the concepts of risks, management of outcomes, and impact on plan liabilities.

Answer   You can't be an actuary if you don't understand statistics. I use it every day at work. Not necessarily the way I learned it in school, but at least the basic principles. It is used in actuarial valuation (with decrement tables, annuities, etc.) and in many other day-to-day actuarial tasks. It is also required for the exams (directly in the first exam and as part of actuarial theory in the others).

Answer The main purpose of actuarial mathematics is to calculate risk, and the only way to do this is through probability and statistics. For an insurer, the only way to figure out how much to charge his customers is by calculating how much they are likely to claim.

Answer Virtually every business problem the life actuary deals with involves the assessment of risk, i.e., the value of a future event contingent on assumed probabilities. Being comfortable with this concept is essential for daily work; actually applying advanced statistical concepts is much less of a requirement, although the opportunities to do so are increasing.

Answer Probabilities are a big part of the first exam and statistics a big part of the third exam. If you chose to work in the CAS field, statistics will certainly be a bigger part of your work and study than in the SOA field.

Answer Probabilities and statistics are the root of actuarial science. They are essential to the actuary and must be mastered. To understand and grasp actuarial mathematics subjects such as life contingencies, it is necessary to grasp the basics of probabilities. In Courses 1 [2002] and 3 [2002], those skills will be tested.

Answer Probabilities of events occurring or not occurring are the backbone of actuarial science. Probability of death, of a car wreck, probability of continued survival. Use of probabilities and statistical distributions can appear on any exam, although mostly in Courses 1 [2002] through 4 [2002] and Course 7 [2002].

Answer Most events and risks evaluated by an actuary are contingencies that can generally be assumed to follow a probability distribution. Actuaries also calculate probabilities (event to occur, having a negative return), expected values (rate of return, age at death, etc.) and volatilities (rate of return, sensitivity of the liabilities).

Answer Just to understand the basics of an actuarial valuation, you need a strong knowledge of probabilities. Once you get that, you have the power to modify the contents of a valuation, or to solve a totally new problem.

Answer Ah, it relates to the calculus question. Everything that contains the word "expected" relates to probability and this is the core of actuarial science. "What is my expected loss or expected profits on this block of business?"

Answer Mortality tables are themselves probability distributions, aren't they? Statistics helps us assess the mathematical validity of the tables by means of confidence intervals, and guides us in determining how much data to collect.

Answer Probability and statistics are again very basic building blocks needed to analyze data, build models, etc., in the projects assigned at work. Again, the syllabus on the mathematical SOA exams is always more technical than the skills you'll ever need in real life.

Answer Some form of probability or statistics is used on a monthly, if not weekly, basis. Examples would include fitting a curve and testing its fit for calculating trends; calculating the probability of an event; calculating the standard deviation of a series of observations, performing a Monte Carlo simulation; etc. Most of this is covered in the first four exams of the CAS.

Answer I work mostly in pension. In that field of practice, probability and statistics are very important since most of the calculations are based on probabilities. Examples: the probability of someone surviving to retirement, the probability of dying a few years after retirement, the probability of someone leaving the workforce before retirement.

Answer Courses 1 [2002] and 4 [2002] have the majority of the statistics problems. It is important to have this background—probably more so in P/C. Credibility of past experience often plays a role, as does frequency and severity. Also, within risk management, stochastic scenarios and the distribution shape are important to consider.

Answer Probability and statistics may be used from time to time on the CAS side to estimate the price of new products (we have no data for those). For instance, in estimating how much a credit card company should charge to provide "delayed baggage" insurance, an actuary could answer the following questions (and then estimate the cost of providing this "coverage"): What is the probability that the baggage be delayed? What is the probability that cardholders with this "coverage" will be aware that they have it, and will then use it? What is the expected value of the loss, and what is the impact on the cost of providing the service of various "limits of coverage"? I find it very hard to relate this to examinations (I wrote them a while ago).

Answer In retirement consultation, very useful for valuating a pension plan.

Answer Probability of decrements.

◆◆◆

Q **Why do actuaries need the theory of interest? Please give examples and relate them to the SOA or CAS examinations.**

Answer Very important for what I am doing, time value of money is a key concept, actuarial present values, rate of return formulas, amortization tables, etc., are all concepts that I have to play with very often in my work even if the way I work with them is different from an examination in Course 2 [2002], for example. Excel is used a lot in playing with these concepts.

Answer This is the basic element of the calculation of today's value of any future payment of one dollar. It is the cornerstone of our field.

Answer It is essential for the calculation of annuities and the understanding of the time value of money. For example, we use it when we calculate things payable at retirement with money accumulated today, or when we want to know what is the value of a pension fund today considering what the membership of the fund might be in the future. Again, it is tested directly in one of the first exams and comes back indirectly in the others.

Answer When dealing with a client, we are looking at the overall result of the company, and this includes investment income, future claims, future revenues, etc. The theory of interest is crucial when it comes time to take those amounts into consideration. It would not be right to use an amount that will be obtained in 10 years, and this is where discounting comes in. The whole point of theory of interest is to calculate the company's financial situation at a certain point in time.

Answer Similar to the previous question, virtually every business problem the life actuary deals with involves the assessment of risk, i.e., the value of a future event contingent on assumed probabilities. The present value of a future event requires the application of the theory of interest.

Answer Theory of interest is the basic [tenet] of many actuarial mathematics and finance concepts. This material teaches you the value of money over time. This has many applications in everyday life and work life. It is also a big part of the second exam.

Answer Present value of annuities.

Answer The theory of interest is needed to understand the basics of actuarial mathematics. The simple concepts of present value and annuities are present, introduced, and explained in detail in the theory of interest, and are everywhere in actuarial science. In Course 2 [2002], these skills are tested.

Answer   Time value of money (i.e., accumulation and discounting) and understanding the basic structure of a bond are very important for calculating reserves, premiums. They are crucial for asset-liability management.
Courses 2 [2002] through 8 [2002] use these concepts.

Answer   Basis for discounting future value of loss, benefits, etc. Also used in
projecting figures in the future.

Answer   That is *the* required course. If you don't understand this one, you may as
well forget an actuarial career.

Answer   The theory of interest is crucial. The time value of money is one of the
underlying principles of the insurance industry, only insurance takes it
one step further by applying statistics.

Answer   Probably not all that necessary now that most work is done on computers
using interest vectors.

Answer   The theory of interest is one of the essential building blocks of actuarial
mathematics. It is needed to define present and future values, for example. Concepts such as calculating present values of bonds and calculating loan payments and outstanding mortgage values involve the theory
of interest.

Answer   Actuaries in property and casualty insurance are constantly discounting
future streams of payments to calculate present values. They also need
to understand annuities since they are sometimes used in the claims settlement process. Beyond this, it is not being used too much.

Answer   In pension, the theory of interest is an important subject. The payment
stream after retirement is based on mortality and interest. It is also
needed to project ahead or discount employee contributions.

Answer   Interest theory and time value of money are extremely important in
any investment-product context (Course 2 [2002] of the SOA exams).
Reviewing cash flows, profitability, and understanding gain/loss scenarios all hinge on the theory of interest. It is particularly important for
actuaries in the investment field.

Answer   CAS only: Present and future value calculations (investment of insurance funds and discounting of loss reserves). Some annuity calculations.

Answer   For valuating pension plans we need the concept of present value.

Answer   Calculation of present value of future stream of payments.

◆◆◆

Q Why do actuaries need the mathematics of finance? Please give examples and relate them to the SOA or CAS examinations.

Answer Finance is a big part of the second [pre-module] exam.

Answer Mathematics of finance is also needed to understand the basics of actuarial mathematics. In Course 2 [2002], an extensive and deep understanding of finance is needed. This knowledge and skill will also be used in the workplace. Often, an actuary will be asked to do some financial analysis. A good basis in the mathematics of finance is necessary for a good actuary.

Answer Actuaries need to understand assets as well as liabilities in order to properly set reserves and premium and dividend rates. Actuaries now need to understand both sides of the balance sheet to do their job correctly. Courses 5 [2002] through 8 [2002] (SOA) really hit on this.

Answer Needed when working on the asset side of a pension plan.

Answer The Course 2 [2002] exam. Also, insurance products relate very closely to the time value of money and finance.

Answer Financial mathematics is used when valuing pension assets. Also, a basic knowledge of financial markets is always useful when dealing with clients and in devising models. Investors and their advisors are becoming more and more informed, leading to more sophisticated market developments, products, and services. As an actuary is, in most cases, at least indirectly affected by financial markets, a basic knowledge of financial mathematics is highly recommended. Course 6 [2002] of the SOA examinations is almost entirely based on financial mathematics. Although probably more technical than most actuaries will ever need, it provides an excellent base.

Answer More and more actuaries are getting involved in the investment side of the business, particularly with DCAT [dynamic capital adequacy testing]. Although not everyone will use it, it is a good idea to be familiar with it in order to be a well-rounded actuary. Theory around cash flow and duration matching are also in common use. I believe this is now being covered in CAS Course 8 [2002].

Answer In pension, you have the promises made to a participant to receive a pension, but you also have the employee and employer contributions that make up the assets. You need to know about investment.

Answer CAS only: Finance is not used *per se* in our day-to-day work. However, knowledge about the effects of diversification may prove to be useful with respect to *planned growth* in P/C. This would relate to actuaries who have a more strategic role—at the executive level, or close to that level. Knowledge about the risks related to various investments (bonds, stocks, etc.) may prove to be useful in discussions at a higher level (executive level). Generally, knowledge about finance is very good to have, although the actual use of it is limited in the day-to-day work. No link with exams.

Answer Finance related to good consulting when valuating liabilities.

Answer Investment science.

◆◆◆

Q Why do actuaries need economics? Please give examples and relate them to the SOA or CAS examinations.

Answer Great background to have for working in retirement or asset consulting so you understand more of what is going on *in the real world*. The only thing sometimes is that economics is a very theoretical science. Sometimes it is difficult to see a real-world application to some theories seen in Course 2 [2002], for example.

Answer To understand the link between the liabilities of a plan and the assets underlying the plan.

Answer A lot of our work depends on finance. For example, with the market situation today, pension funds are losing money. This fact should guide actuaries when they give advice to their clients on when to file an evaluation or the decision to improve the plan, for example. It is tested in the examinations in Courses 2 [2002] and 3 [2002].

Answer Economics is a big part of the second exam.

Answer Economics are also needed to understand the basics of actuarial mathematics. In Course 2 [2002], competency in economics is tested. This knowledge and skill will also be used in the workplace and serve to understand the ways a company and the market work.

Answer Actuaries need to be able to understand the structure and the workings of the different investment markets in order to manage their assets that back their liabilities well. Course 2 [2002] and Courses 5 [2002] through 8 [2002] touch on this.

Answer Set appropriate economic assumptions for actuarial valuation: discount rate, rate of return on assets, etc.

Answer The thing I remember about my economics class is the *marginal cost* theory, which I apply very often. But I'm not sure if I needed this class.

Answer Course 2 [2002]. Also, the ideas of balance sheets are crucial even for pension plans and for the reserves of an insurance company. Pension actuaries must weigh the assets and liabilities of a pension plan against each other.

Answer Actuaries should have some idea of how macroeconomic events in the economy may affect the sectors of the economy that have an impact on their business. For example, how will a slowdown in inflation affect long-term interest rates?

Answer Actuaries need economics since almost all assumptions are based on current economic market conditions with projections for future economic outcomes.

Answer Some economics concepts can be used in modeling, for a better understanding of the impact of rate changes, for example. Simple concepts such as the law of supply and demand.

Answer CAS only: Economics is not needed in our day-to-day work. Knowledge about it may certainly come in handy from time to time, but then again, more at a higher level (executive level).

Answer Depending on the field, it is not always necessary.

Answer Economic knowledge is needed to try to understand the needs of clients.

♦♦♦

Q Why do actuaries need risk theory? Please give examples and relate them to the SOA or CAS examinations.

Answer Risk theory helps me understand the foundation of actuarial science. It is very important, I think, to be strong in this technical area since the ideas involved come up on a daily basis.

Answer Basic to our job is managing the risk related to a plan.

Answer I rarely use risk theory at my level. But it is important for the examinations.

Answer Course 3 [2002] tests these skills.

Answer Actuaries are trained to put a value on risk and handle future contingent events. Risk theory is the real fundamental bridge between life contingency theory and the business of insurance. Courses 3 [2002], 5 [2002], and 8 [2002].

Answer   Understand risks, model risks to eventually put a value and cost on it.

Answer   Risk theory is the heart of actuarial work. An actuary is an expert in the assessment and management of risk.

Answer   Risk theory is the basic building block of the P/C business. However, as indicated earlier, the level of sophistication is rather lacking in the Canadian marketplace. However, it is helpful to understand risk theory in order to perform the daily work of a P/C actuary.

Answer   CAS only: This is the basis of the pricing work in P/C. I cannot say, however, that what I learned in school with respect to risk theory helped me in my work.

♦♦♦

Q   **Why do actuaries need loss modeling? Please give examples and relate them to the SOA or CAS examinations.**

Answer   I guess it is very important in CAS, but is less important in fields such as asset consulting.

Answer   I am not yet familiar with loss modeling.

Answer   I think this is more of a CAS thing or perhaps also a reinsurance thing. You need to be able to calculate the probabilities of incurring a loss before you can accurately set a price for an insurance premium. Loss modeling comes up in Course 4 [2002].

Answer   CAS stuff. Used in pricing products by modeling future expected losses. Needed since non-life risks generally have the following characteristics: time of event unknown (so need a frequency distribution) and size of loss unknown (so you need a loss distribution).

Answer   More useful for CAS, I think.

Answer   Loss modeling is a fairly useful tool that is hardly ever used, at least in my experience. Lack of size (and therefore lack of data) is one of the problems encountered when trying to do loss modeling. Often a lack of time and resources will also force a company to use a broad-brush approach in its pricing and reserving modeling.

Answer   CAS only: Loss modeling may be used to forecast the severity of certain events, and also to determine how variable results will be from one event to the next (link with credibility of results). For example, in looking at automobile theft, vandalism, and fire, loss modeling may be used to determine the shape of the curve that best describes severity (average cost). Once this is done, one can determine how variable this severity will be, and therefore how many observations are required in order to get credible estimates.

◆◆◆

Q What stochastic ideas and techniques do actuaries use? Please give examples and relate them to the SOA or CAS examinations.

Answer I know areas of the actuarial field where it is extensively used and is important. This is not yet the case in asset consulting (at my level). But I know that stochastic ideas are very important in asset and liabilities management, an area I would love to get into later on in my career.

Answer The only method used frequently is the Monte Carlo simulation, mostly for the projection of the assets of the plan.

Answer Continuous Markov chains are used by actuaries and are tested in Course 3 [2002], I believe.

Answer Becoming more prevalent, especially with modeling possible future interest rate patterns when determining reserve amounts for life insurance and annuities. Also used for sensitivity testing and pricing of minimum guaranteed death benefits for segregated funds. Course 8 [2002] had a big section on this. Course 7 [2002] Pre-test had this.

Answer Projections of pension plan assets or surplus based on stochastic distribution of future interest rates. You can then determine the future distribution of values by percentile, calculate the probability of having a value less than some fixed amount, etc.

Answer To forecast what are best and worst case scenarios under different sets of hypotheses for surplus or deficit in a pension plan.

Answer Stochastic modeling of the cost of face amount guarantees on segregated funds.

Answer CAS only: Stochastic techniques are not widely used in Canada. They may be used in the area of DCAT [dynamic capital adequacy testing], although I do not know anyone who has programmed or is using a stochastic model in Canada to do DCATs.

Answer More in asset consulting than liability consulting.

◆◆◆

1.6   Complementary Disciplines

In addition to being good in mathematics, economics, and other scientific subjects, actuaries need to have a broad arsenal of other skills. What college and university courses should they choose to acquire these skills?

Q   **Which are the most important complementary disciplines for an actuary and why?**

Answer   Finance and accounting. The actuarial profession, especially in pensions consulting, is increasingly exposed to managing and considering the asset side of the balance sheet. A well-rounded professional therefore should have exposure and an understanding of accounting requirements and financial opportunities related to pension asset investments.

Answer   Software skills (not really programming, but a good working knowledge of Microsoft Office is key). Interpersonal skills, communication skills (used every day in a working environment), understanding of financial markets and economics (it's my job!).

Answer   Finance and economics. The pension industry is driven by the assets behind pension plans.

Answer   I would say two: finance and computer science. It's essential to know about finance because everything that we do is related to finance. We use annuities, rates, present values, and so on. Understanding the value of money over time is essential. Also, because it would be too complicated to calculate everything by hand, we use computers a lot. I've seen someone with a Master's degree not getting a job because he couldn't work with Excel. Computers help us do our work faster.

Answer   Statistics and probability, risk theory. Everything relates to this.

Answer   For a pension actuary: I think accounting is becoming increasingly important in a consultant's job. Companies (especially the larger ones) are more concerned with the annual pension expense and it is key that actuaries have a good accounting background. A strong understanding of financial concepts is also very important (these courses are also useful for the later SOA Exams 6 [2002] and 8 [2002]). Programming courses (Microsoft Visual Basic) are useful for beginning actuaries.

Answer   For the first three actuarial exams, finance and economics are helpful. During my work in a casualty insurance company, I used a lot of programming skills (SAS and Microsoft Visual Basic). These skills are used to extract data from huge databases to do calculations. During my work in consulting (pensions), I used some accounting and language skills. In preparing reports, both skills were useful.

Answer Finance: knowledge of balance sheets and understanding the impact of our work in the real life!

Answer I would say that business courses are necessary to complement a good actuarial program. Courses such as finance, economics, management and marketing are essential for actuarial students. Communication courses, as well as language courses could be an asset. Computer science courses are also very important.

Answer Computer science—work with computers every day and it's not all programming; investment and economics—much of an actuary's job is understanding how investment markets affect the risk assumed by an insurance company; medicine—used in underwriting business; law—understand contract law and legal regulations; communications— both written and verbal are important, especially when dealing with auditors; marketing—have to convince people to buy your products or services.

Answer Chartered Financial Analyst: specialist of the asset side of pension plans. Micro/macro economics: to determine appropriate economic assumptions. Communication: for presentations, understandability, and so on. Computer programming, etc.

Answer As an actuary just starting out, I find programming skills extremely valuable. Most junior actuaries will be required to program in a variety of languages. Having good software skills in general will always make for a more efficient actuarial analyst. Other disciplines that can be useful are economics and finance. As an actuary's career develops, softer skills such as management, delegation, verbal and written communication, and relationship building can play an important role as well. However, actuarial students rarely consider them essential when they are still in school.

Answer Computer science early in your career because that is what you do. Then business or law should be helpful.

Answer Strong software skills are necessary. Microsoft Excel and PowerPoint. Programming is also necessary. These are all tools used in doing the work. Doing calculations and finding results require broad and technical understanding of economics and finance. We have seen this year the negative impact the stock market downturn has had on the assets of a pension plan. It is important for an actuary to have a feel for the economy and how it is related to his work.

Answer   Finance and economics: Useful for seeing the *big picture* and broader context of the actuarial field. Ideas from finance and economics are required when making assumptions about future pricing of actuarial products. Computer science: You need to know how to program. Period.

Answer   Actuaries have such versatile training and have a lot of competencies so that they can excel in a variety of work—finance/risk/insurance related.

Answer   Liberal arts, because you can learn all of your actuarial skills on a company-supported program of self-study, but we will not pay a dime for philosophy, linguistics, or novels of the 19th century. A well-rounded actuary is more valuable to us in the long term than one who has had a narrow technical education.

Answer   Accounting and finance for obvious reasons (as you go up in the hierarchy of an organization, it becomes really important). Management and human skills—too many actuaries lack these skills (and still become manager because of their professional status, e.g., Fellow).

Answer   Economics: to project future economic scenarios/be familiar with current economical situations; finance and accounting.

Answer   It would be finance, economics, management, and computer science. This view is based on my experience as a CAS corporate actuary. However, I believe the same would be true for SOA or for a pricing actuary. Finance is important because when dealing with the accounting department, you must be able to speak their language. Proper understanding of financial statements is also important in a wide variety of situations (not just in your own company, but also, for example, for the pricing of some insurance products for corporate clients). Finance includes an understanding of investments, an area in which more and more actuaries are getting involved. Economics is important to understand the concept of supply and demand in order to make good pricing decisions. If an actuary wants to climb the corporate ladder, then management skills are invaluable (not only in order to manage your employees, but also to meet your boss's expectations). In addition, all actuaries have to be able to do some form of programming at one point in their career.

Answer   Accounting and finance are very useful for valuation type work in the life insurance industry.

Answer   Software skills, French and/or English (verbal and written), accounting, economics.

Answer   Finance: To know about assets and liabilities. To understand a client's needs.

Answer Accounting: the knowledge of accounting concepts is crucial when using an insurance company's annual statement. Finance: financial theories are often applied to actuarial concepts and widely used in actuarial calculations (e.g., discounting of liabilities). It is also important for an actuary to understand the investment portfolio of an insurance company. Economics: principles of economics can be used to solve actuarial problems (e.g., analyzing supply and demand curves to price insurance premiums). Management: actuaries eventually get to a level in their career where they have to supervise and manage others.

Answer Communication skills cannot be overemphasized! Actuaries have always had a well-deserved reputation as "bright guys everyone wants to avoid." An actuary's ability to solve a problem is worthless unless he can persuade a non-actuarial audience that the solution substantially meets everyone's needs. Computer literacy is also important because almost all complex actuarial problems are solved with computers in today's environment.

Answer Computer science, because all of the more junior positions require programming skills.

Answer Investments and economics: the complement of liabilities is assets.

♦♦♦

Software Skills

Here is what the survey respondents said about the importance of computer skills in general:

Q What software skills should actuaries have and why? Please give examples.

Answer All Microsoft Office tools (especially Microsoft Excel), databases (often firms have their own database system that is learned on the job), and logic. Knowledge of time management software is a must to effectively manage time in and out of the office!

Answer Skills are more related to problem-solving approach than real programming skills.

Answer Knowing all Microsoft tools such as Word, PowerPoint, Excel, etc., well. Being comfortable with searching for information on the Internet.

Answer Actuaries mostly use Microsoft Excel and should feel comfortable using it. Since we are playing with numbers all day long, any software that performs similar operations can be used.

Answer Programming skills are needed. Also, Microsoft Excel is a commonly used tool and the actuary should be very comfortable with using formulas and editing data. Sometimes Microsoft Access is used for data modification or verification.

Answer Strong Microsoft Excel skills are required, I think, in every company.

Answer Microsoft Excel, Access, and Visual Basic programming.

Answer Microsoft Excel is a definite must. It is used in the day-to-day routine of an actuary. An actuary should also have good computer programming skills and be comfortable with the Internet. Like in many careers, the computer is one of the basic tools of the actuary.

Answer *Spreadsheets:* lots of work is done on spreadsheets these days, instead of programming. If you can build effective spreadsheets you will save a lot of time and look better in front of your supervisor, who will be able to check your work more easily.
Databases: they are about the concepts of fields, items, relations between tables, and allow you to work efficiently.
Word processing and touch-typing: not key skills, but you will do a lot of typing over the career, so why not learn it? Macro writing, in Microsoft Visual Basic, for example. This can make your life a lot easier when it comes to repetitive tasks.
Programming: this skill is required for particular jobs such as obtaining data from mainframe systems. And probably the most important skill: knowing which tool to use in a given situation.

Answer Microsoft Office or equivalent, Excel (including macros), Microsoft Access, GGY's AXIS, some basic programming skills, some computer operating system skills, mainframe computing experience. Most spreadsheet work is done with Excel. In addition, Access is used for data queries to get results for subsets of data. AXIS is an actuarial pricing and valuation tool. Mainframe computing is required in big companies with large blocks of business. Many companies also still use in-house programs created in APL.

Answer I believe that Microsoft Visual Basic for developing Excel macros is a must.

Answer Microsoft Excel and Word, and the ability to learn quickly.

Answer Excellent knowledge of Microsoft Excel (how to be efficient with it), Microsoft Access. Be good at programming languages—this differs from company to company. Some use SAS, Fortran, and Microsoft Visual Basic. It depends. If you have a good basis in one, your brain is already able to think in a programming environment.

Answer Familiarity with Microsoft Office (Excel, Word). This is the industry standard. All of this can be picked up on the job. As a new recruit you will look good if your ability quickly becomes a center of knowledge for the unit, so it pays to deepen your expertise.

Answer Since nowadays most work is done on computers, software skills are a must. Spreadsheets are extremely important. Word processing software and presentation software are also important. Most companies have their in-house programs and software with which one must become familiar.

Answer I think everyone should have software skills nowadays. Actuaries, more specifically, need Microsoft Excel and Access skills (or more generally spreadsheet and database skills). Actuaries typically work with a lot of data. Even if extracting data via a programming language, the quantity may still require some processing through a database to make the information more manageable for analysis. Databases, however, are not suited for analysis (at least, in my opinion). One does require in-depth spreadsheet knowledge to perform regression analysis, Monte Carlo simulation, Bayesian estimate, etc.

Answer Actuaries should have good Microsoft Office skills, especially Excel, Access, and Word. They should have strong programming skills as well. Entry-level jobs, in particular, require good software skills.

- *E-mail.* Capacity to use e-mail (obvious, I guess).
- *Excel.* Capacity to use Excel (most of the calculations are done in Excel).
- *Word.* Word processing: to write memos/documents (speed of typing is important).

To be an actuary, one must like computers because about 90% to 95% of the work hours are spent on a computer.

Answer Good in Microsoft Excel, Word, and Access.

- *Excel.* Advanced knowledge of Excel is required since most companies use Excel to put together actuarial analyses.

- *Access.* At least intermediate knowledge of Access is required. The actuary who is able to run complex queries will generate better data as a basis for actuarial analyses.
- *Word.* Basic knowledge of Word is required to convey results and findings of actuarial analyses.
- *PowerPoint.* Basic knowledge of PowerPoint is required to prepare presentations to management or clients.

Answer Capability to program because most junior positions require programming.

Answer Microsoft Excel: tables, charts, very powerful.

◆◆◆

Programming Skills

Here is what the survey respondents said about the importance of programming skills in particular:

Q **Which programming languages do actuaries need and why? Please give examples.**

Answer Rarely. I need to write macros in Excel. That is about it.

Answer None. All companies have their own software now.

Answer Usually, each actuarial firm as its own program, so I don't think there is some particular programming languages needed. Of course, Microsoft Excel and Visual Basic are used a lot. I would say that an actuary should know at least one of the common programming languages (Fortran, C++, etc.). With that knowledge, it should be enough to adapt to others.

Answer In my day-to-day work, I use Microsoft Visual Basic (for macros) from time to time. Besides that, it is mostly company-specific programs. Therefore, more than knowing a single language inside out, I believe it is more important to have a strong understanding of programming methodology.

Answer I used SAS in a casualty insurance company and in government, and Microsoft Visual Basic in all of my internships. Although I took C++ courses at university, I have never used this programming language.

Answer I don't use any language, but the logic behind it is used for company-specialized software for actuarial valuations.

Answer Microsoft Visual Basic and Visual Basic for Applications are often used in the field. Knowing how to program macros and use them is often a great advantage. APL, although now more and more scarce, is also a programming language that has benefits since it is still used in some insurance companies.

Answer Experience with any programming language for *real* applications (not just basic applications) is very important. It is important not just to program, but also to go back and change programs, both yours and those written by others. Such experience promotes logical thinking, modules, testing in units, good documentation, patience, etc. Good languages to learn are those with structure (e.g., Microsoft Visual Basic, Fortran) and APL (because it is so different, and helps you think of matrices).

Answer Actuarial tasks are becoming less and less programming oriented, now that programs such as AXIS do many actuarial calculations. APL and Microsoft Visual Basic are probably mostly used. Maybe some SQL.

Answer Basic knowledge of programming languages such as Microsoft Visual Basic or Fortran is necessary to perform valuations in an efficient way.

Answer Although the programming languages that are used vary by company, some of the more common ones that I have encountered are APL, SAS, Focus, and Microsoft Visual Basic.

Answer Programming was useful at the beginning of my career, but seldom used today. Logical thinking is required, and then you have to be able to coach a *real* programmer towards what you want. You don't do it yourself anymore.

Answer The ability to learn a programming language quickly. I have seen many languages at work.

Answer It depends on the company—a lot of companies have in-house software. We use Fortran, Access, and AXIS. In P/C insurance, SAS is often used. A good knowledge of Microsoft Excel is always an asset.

Answer Very few. Familiarity with Microsoft Excel is useful. Some areas still use APL.

Answer APL, SAS, and Microsoft Visual Basic.

Answer I would say that programming skills are an asset but not a necessity. I personally hate programming and have avoided it since birth.

Answer The programming languages used by companies can vary. But actuaries, at the entry level especially, need programming skills in order to retrieve data for analysis. Very few companies have programmers involved in what amounts to data mining. Most IT [information technology] programmers are business programmers, worrying about transactions. Basically, they worry about the transaction of writing a new policy or paying a claim (financial information), but do not understand actuarial concepts such as earning premium or accident year data.

With the proliferation of databases, programming languages will continue to flourish. Good Microsoft Visual Basic or SQL query skills will be required to extract and work with information. One does not want to repeat a series of manual commands every month, writing a macro is much more efficient.

A programming language in common use is SAS. Some companies also still use APL, although this is diminishing. Finally, some understanding of basic programming concepts and languages (such as Cobol) can help, for example, when you are working with the IT staff trying to debug a rating system.

Answer   SAS (manipulation and analysis of large data sets, generating actuarial reports), APL (for reserves calculations), and Microsoft Visual Basic.

Answer   Actuarial students should have good programming skills. However, I don't believe it is crucial that students know one language over the other. If students have good programming skills, they will be able to learn the language used by the company they are working for relatively quickly. I do believe that being able to use Microsoft Visual Basic in Excel is an asset. Our company still uses Fortran, although knowing Fortran is not a requirement to get a job at our company. However, having good programming skills are.

Answer   I think most actuaries can learn programming on the job. Knowing APL, C++, etc., is not mandatory—in some roles they're not even used! Now profit testing programs like TAS or MoSes are being used more.

CAS only: SAS—this is used by most insurance companies and consulting firms to extract and manipulate data. *(This is a must.)* APL— less used today than in the past, but still used by reserving actuaries. Microsoft Excel (including Visual Basic for Applications)—90% of work is done in Microsoft Excel, with the exception of data extraction (this is not a programming language, but I thought I should mention it anyway).

Answer   Fortran and valuation programs.

Answer   Microsoft Visual Basic and SAS.

Answer APL, Microsoft Visual Basic, and Visual C++. Microsoft Excel and database knowledge.

◆◆◆

Business Skills

It is often said that good business skills are essential to succeed in the actuarial world. Many companies now specialize in the teaching of business skills. It is also said that you need luck to succeed in life. But what is luck when it comes to an actuarial career? Seneca, the Roman philosopher, has provided us with a definition of luck that is quoted by many business schools today: *Luck is when preparation meets opportunity.*

Preparation in the context of an actuarial career is fairly easy to define. Becoming an Associate and then a Fellow of one of the actuarial societies comes close. However, when I conducted the interviews for this guide and asked many actuaries what they considered to be one of the most important skills required to be a successful actuary, they almost unanimously put communication skills at the top of the list.

This prompted me to browse the Internet and find out if there were some common elements among the many popular lists of essential skills to success in business (think actuarial career). My query scored over 200,000 hits. As you can imagine, most hits were irrelevant. But some skills came up in most of the lists, not necessarily in the following order:

1. Networking skills
2. Communication skills
3. Research skills
4. Team-work skills
5. Critical thinking skills
6. Management skills
7. Financial skills
8. Presentation skills
9. Decision-making skills
10. Leadership skills

A few other skills were cited that you don't normally think of, such as the ability to relax, the ability to forgive and forget, the ability to sell yourself, and the ability to control your emotions.

Try to keep this list in mind when you are reading through this guide and are getting overwhelmed by the long lists of technical requirements. Yes, most actuaries are good in math and most have many of the qualities considered essential. But almost all successful actuaries have learned to survive under enormous pressure, the pressure to succeed in examinations while working full-time and the realization that, as one actuary put it to me, you don't have to ace the examinations, all you need to do is pass them. Keep this in mind as you enjoy the challenges presented in this guide.

Many if not most of these skills can be learned. Business schools specialize in developing these and related skills and are discussed on the websites of most business schools.

Q **What business knowledge and skills do actuaries need and why? Please give examples.**

Answer I am not sure I understand the question, but ultimately, selling skills and defending ideas can be great when meeting with clients.

Answer Understanding a client's business, including financial statements, calculations of profits, etc.

Answer At a higher level, actuaries sell services to clients. So actuaries need to be good in persuasion, understanding needs, and foresee problems or requests. Honesty is also very important. They need to be aware of the market in general. To understand their clients better, they also need to check specific fields in particular (if your client is a factory, you should know if the market is good for that field, not just for your client or in general). Actuaries also get to manage clients' teams: prices to charge, tasks to perform, who's to work together, time allocated to a project, and so on.

Answer A good background in business is necessary to an actuary. A knowledge of finance is essential in the study of actuarial mathematics, but even skills and knowledge in marketing and accounting can be useful since you will often find actuaries in the marketing and assets and liabilities management department of an insurance company. Since actuaries often hold management positions, management skills can be useful.

Answer There are others, but I would start with basic accounting (balance sheet, income statement, double-entry accounting) and finance (investments/ assets characteristics).

Answer Knowledge of investment markets and types of investments, economics, insurance product structure, knowledge of how interest rates affect insurance liabilities and assets. Values of assets affect the amount of surplus that an insurance company has, which limits how much new business they can write. Values of liabilities affect reserves and surplus as well. There's plenty more, but that's all I can think of right now.

Answer Listening. Ability to solve problems.

Answer Understanding of accounting principles and financial statements, because you are sometimes responsible for big amounts on these reports.

Answer Ability to make decisions. This is often weak in actuaries, because they by nature see many sides to a discussion, and can accept many right answers. A bad decision taken is better than several good ones deferred.

Answer Organizational and time management skills are important—one must often juggle many projects at once with strict deadlines. Project management skills—knowing how to initiate, do, review, and close a project with many variables and constraints, deadlines, and objectives. Communication skills—verbal and written skills a must. Professional ethics—integrity, treating others with dignity and respect. Professionalism—a sense of professionalism in style, presentation, and communication to others, verbally and in writing.

Answer This depends on the ambition of the person involved. Generally, the more ambitious, the more business knowledge and skills are required. Actuaries who are happy working in the back room and are not interacting with people other than their manager and coworkers probably don't need too many business skills. However, anyone who wants to climb the corporate ladder requires business skills. Indeed, let us not forget that this is what we are doing: running a business. The best actuaries in the field are, first and foremost, businessmen. They can understand the difference between an actuarial indication and the price the market will bear. They understand the implication on the company of their decision with regard to IBNR [incurred but not reported] loss reserves or reinsurance. They get involved in projects and understand the work flow of the organization, the difference between the bells and whistles, and necessary system enhancements. Business skills required include economics, marketing, management (both personal, time, and project), finance, investment, and communication.

◆◆◆

Communication Skills

Here is what the survey respondents said about the importance of communication skills as their careers unfolded:

Q What communication skills do actuaries need and why? Please give examples.

Answer The more skilled actuaries are, the better they are, as I have found out since working full-time. Especially in Montreal, being able to speak and communicate fluently in both English and French is a *great* asset. For a junior consultant, it is of the utmost importance to communicate very well with the seniors so we understand exactly the work that needs to be done and once completed, to be able to explain it to the consultant in clear words. Listening is also a forgotten skill, but very important in day-to-day work. Presentation skills become more and more important, I assume, as you grow in the business and have to meet with clients and present them ideas and reports. Being able to support your ideas and organizing your thoughts are also key skills.

Answer Clarity, since it is difficult. Simplicity, since the client must understand.

Answer Knowing at least two languages, enough to be able to communicate, is essential. It is not unusual to encounter French-speaking clients, for example. Canada is so bilingual that it's not an option anymore. Also, an actuary needs to be able to express his thoughts and his knowledge. It will happen often that a more advanced actuary needs to explain something to a new one or even to a client. So being able to be clear, not too complicated and see when the other person doesn't understand is essential. The same skill applies for writing (the annual statements, for example, need to be clear, but simple).

Answer You need verbal skills to give presentations to your colleagues and to clients (particularly in the consulting field). Your writing skills will be useful to write actuarial evaluation reports or prepare internal status documents.

Answer Very good communication skills in order to gain credibility from people we are working with and to explain simply what we have done and why.

Answer Presentation skills are necessary since it is often required from actuaries to present their research results, projects, or recommendations. Actuaries must also be able to *sell* an idea. In the consulting business, actuaries will interact with clients and need to be able convince the client of the necessity of a benefit plan for example. Actuaries also sometimes need to explain their results and recommendations. Hence communication skills are, as much as mathematics and business skills, essential in the making of a great actuary.

Answer Verbal: speaking to other actuaries in technical language, speaking to non-actuaries in nontechnical language, presenting to management/board on reserves (appointed actuaries), presenting updated pricing models to underwriters, leaving phone messages. Written: documenting in clear and understandable language, writing some letters and reports (especially in consulting), e-mail. Listening: gathering information, learning about other areas of the company, learning other peoples' terms so you can speak to them in their language.

Answer Written: you often need to write important reports for management and regulators and/or auditors. Internal documentation of processes is needed as well. You also need to use e-mail effectively to communicate with non-actuarial staff, the field force, and customers who need to have technical concepts explained in nontechnical language. Verbal: same reasons as above, without the written reports for regulators and auditors. Public speaking: you are often required to make presentations to audiences with varying actuarial knowledge.

Answer You need good writing and verbal skills. Mastering two or more languages is a must.

Answer The biggest challenge is to understand what you are doing and then to be able to explain it to people who don't have an actuarial background. Therefore, it requires excellent communication skills if you don't want to spend your life in front of your computer.

Answer Expression. It might often be hard to express a mathematical calculation in words, but this is a necessary ability. When working with a team, it is necessary to be able to discuss one's work and the need for certain calculations. It is important to be able to give oral presentations and to be able to speak in front of a crowd. In a corporation, you must speak in front of a group to share knowledge and ideas.

Answer You need to be able to communicate complex actuarial concepts to a variety of audiences and levels of understanding.

Answer Verbal and presentation skills to be able to present and *sell* your ideas and concepts to management. This is often critical when you work closely with upper management (such as corporate actuaries).

Answer Depending on the actuarial field, various levels of communication skills are required. For consulting actuaries, communication skills are extremely important. The level of knowledge of clients is quite broad; ranging from clients who are well informed to clients who have only a basic knowledge and depend on consultants to provide them with the required knowledge and information. A consulting actuary must therefore be able to communicate technical information in laymens terms and be able to tailor the information based on the level of knowledge of the clients.

Answer Actuaries need both written and oral communication skills, especially as more responsibilities are assigned to them. Actuarial mathematics is a difficult concept, and it is difficult to explain to a lay audience. Furthermore, at least with P/C companies, actuaries interact a lot with marketing, sales, and branch managers. In addition, all corporate actuaries interact with finance and upper management, as well as information technology. Rarely do actuaries price a product in a vacuum. The actuarial indication is only the beginning of the process. What good is it to price a product at the actuarially sound rate if nobody is going to buy it? Especially if the high price is driven from conservative assumptions. Pricing actuaries often have to explain or sell their recommended increases to a variety of people. When involved in various projects (or business), actuaries are often experts relied upon to help shape the requirements of the project. Good writing skills are essential in these instances. Corporate actuaries often need to explain their IBNR [incurred but not reported] loss reserve calculations to upper management. When a change in IBNR can erase the entire profit for a given year, not only are good communication skills necessary, political savvy is also essential! At a certain level, appointed actuaries also have to report to the Board of Directors. To avoid the "glassy eye" syndrome, good verbal communication skills are again essential.

Answer Actuaries can be faced with solving problems and giving answers to people within the company who are not necessarily mathematically inclined. Therefore, they need to be able to communicate and explain results in a simple way. Communication skills and the ability to explain technical terms and procedures are needed.

Answer 1. Capacity to explain complex concepts with simple words to non-actuaries (*extremely* important).

   2. Listening skills are very important.

   3. Speaking French is a big plus because of Quebec.

   4. Written communication skills are very important (capacity to explain in writing complex concepts to non-actuaries, writing without mistakes).

   5. Public speaking (presenting results to a group a people in such a way that the audience understands).

6. Being able to use non-verbal ways of communicating (visual aids for example).

Answer Written and oral skills are very important because there are a lot of reports to write and presentations to clients and employees.

Answer Communications skills are needed to interact with clients and other personnel.

♦♦♦

1.7 **Actuaries of the Future**

One of the exciting aspects of an actuarial career is the continuously changing nature of the profession. As the world changes, so does the role of actuaries in it. What will be some of the future skills required to adapt to this change? How do actuaries view the need for new skills as the profession evolves? In his introduction to actuarial modeling (see [12]), for example, Jones mentions several active fields of mathematical research such as neural networks, fuzzy logic, and chaos theory as sources for new actuarial techniques. SOA and CAS now actually publish a joint newsletter called *The Future Actuary* (see [22]), providing regular glimpses into advances in the actuarial world.

In some countries, more reliable statistical data, including mortality data, are now being actively collected. Over time, this new information will find its way into the actuarial world. New skills will be required to create credible predictive models based on these data. Knowledge management and global communications will continue to advance. International business will thrive and the need for foreign language skills will increase as a result. The globalization of manufacturing will create a need for new forms of insurance. The fusion of banking, insurance, and wealth management will create new actuarial challenges and opportunities. The list goes on. Equipped with the appropriate skills to tackle these challenges, actuaries will play a key role in assessing the risks inherent in these changes.

Here is how some of the respondents to the survey see the impact of these and other changes on the future of the actuarial profession:

Q **What changes in the knowledge, skills, and mathematical techniques in actuarial practice do you envisage in the next 5/10/20 years?**

Answer More software skills, more interpersonal and communication skills. More time management skills, since workload is getting heavier and heavier since machines can do work faster and faster. Fewer pure mathematics skills.

Answer More communication, more business/finance oriented.

Answer Forecasting techniques are becoming more widespread. Future actuaries will need to become more comfortable with liability projections and integrating future asset forecasts. Again I think accounting is becoming more and more important. The SOA examination in Course 8 [2002] has a greater focus on pension expense and balance liability than ever before. Also, more clients are becoming concerned with the effect that the pension plan is having on the company's books.

Answer I believe that in the future, the emphasis will be more on the refining of the business skills and communication skills of a future actuary. More and more, the importance of these skills is growing and makes the difference between a good and a great actuary.

Answer The mathematics probably won't change, but the regulatory and computer knowledge will greatly change. There hasn't been a major new insurance product type introduced for about twenty to twenty-five years, so if one is introduced, then knowledge about how insurance products are priced and valued will change. Increased regulation and improved computer systems will allow risk calculation to become more fine and, hopefully, more accurate.

Answer Perhaps more Internet-oriented for basic information: services that are perceived as not adding value should be made available to clients on the Internet. We may be concentrating more on how to add value.

Answer More applications of fuzzy logic.

Answer CAS: In the next five to twenty years, actuaries will need to become more efficient as well as more refined in their approaches. However, the top priority will be to improve communication skills. *This is the single most important area for success!* Knowledge and skills: Actuaries will need to be more *non-actuary-friendly*. In the past, actuaries were thought of being in an ivory tower, but times have changed. More and more, actuaries are involved with a number of individuals in discussions leading to important business decisions. In the past, either the actuaries were not involved, or they would be one of the very few parties making the decision with the President of the company. This points towards the need for actuaries to be team players more than ever, while at the same time being able to influence people to make the right decisions. As a result, communication skills will be key to the success of actuaries in the future. Mathematical techniques: More specifically with respect to pricing in P/C, I foresee a much greater use of generalized linear models. With respect to dynamic capital adequacy testing, I would like to see a greater use of stochastic models as opposed to deterministic models. Will it happen? I do not know.

♦♦♦

1.8 SOA versus CAS

Q What is the main difference between working in SOA and CAS?

Answer SOA is a bit less technical and more regulations-driven than CAS. In SOA, you are dealing less with figures, and there is more people interaction (but I don't know CAS well enough to have a perfect answer to this question).

Answer International career opportunities, and the type of work itself.

Answer I believe the main difference is diversification. In SOA, actuaries tend to specialize more into one area, whereas in CAS, actuaries need to be more knowledgeable in several fields. There are advantages to both, and they fit two different types of people.

Answer On the CAS side, work is much more technical. You will have to calculate and to program a lot more. On the SOA side, you will have to work more with people outside of the actuarial profession. You will have to do reports and to be in touch with clients.

Answer SOA will generally deal with subjects more familiar to fresh out of school students. SOA deals more with life insurances and pension. CAS is in itself more general and will require a more broad knowledge of the field because there are so many options out there. It is also more competitive and therefore more studies of the market will be done.

Answer As far as I know, the main difference has to do with whether you are interested about life-related stuff (life insurance, annuities, investment products, pension plans, welfare benefits other than pensions, etc.) or non–life-related stuff (home insurance, car insurance, risk insurance, etc.). Also, it seems that consulting actuaries are put in the SOA group since most work on employee benefits–related issues, which are considered as life aspects.

Answer Although I have never worked in an SOA-related job, here is an extremely simplistic answer. Most CAS-related jobs involve the pricing of automobile, property, and liability insurance products. For example: personal car insurance, home insurance, small commercial businesses, liability coverage for doctors or lawyers, hole-in-one contests at golf tournaments, just to name a few. A main function of SOA actuaries is to price products that are more related to life insurance and pension plans.

Answer Good question, and difficult to answer since I've only worked in the CAS world (except two work terms, but one cannot get a good idea of SOA work from a few work terms). Off the top of my head, I would say that CAS work tends to be less technical than SOA. It seems to me that SOA has a formula for everything (e.g., pricing life annuity), where most of the work is spent refining assumptions. On the CAS front, there is no accepted formula for pricing, just a basic methodology. Pricing work consists mostly of gathering data and projecting it into the future. Pricing in CAS is also very much dependent on competitive market forces. Again, the same is true of reserving, the setting of IBNR (incurred but not reported) loss reserves being mostly educated guesswork on the CAS side. On the environment side, CAS companies are generally smaller (especially in Canada), which would indicate greater opportunities for promotion. However, talented individuals will be able to climb the corporate ladder irrespective of their choice.

Answer I have never worked in P/C, so I would guess product line and the length of the liabilities from what I've heard.

Answer Only 9% of actuaries in Canada work in CAS. In my opinion, CAS work is less repetitive in the sense that you are more bound to be confronted with new problems than on the SOA side. This is the result of a few factors: The P/C industry is extremely competitive in Canada (insurers are always trying to find new ways to be more profitable and gain market share). P/C work deals with two random variables instead of just one on the SOA side. Pricing in P/C is based on 10 to 15 variables, which increases the complexity of the work of CAS actuaries. Because fewer actuaries work in CAS, smaller groups of CAS actuaries work in a given firm (in general, it may not be true in all cases), which leads to a *small-family* mentality. I am not sure about the SOA side, but the demand on the CAS side is currently increasing quickly and is bound to increase even more in the future. To the extent that this may not the case on the SOA side, this could be another difference between the two career paths. Life insurers are profitable, P/C are not, or at least not as much. This is a major difference between SOA and CAS.

Answer Working in SOA (pension field) involves a lot of data manipulation at the entry level and for a few years afterward. In the CAS environment, actuarial analysts get more involved in the analyses (including data manipulation, of course) right from the beginning and get exposed to a better variety of projects.

Answer I can say from experience that the underlying actuarial principles are the same. The main difference is in the minutiae of the regulations and how these have affected the practices and procedures of the industries.

◆◆◆

1.9 Actuarial Accreditation

Q How important is the number of actuarial examinations passed for an actuarial career? Illustrate your answer with examples of careers in companies you have worked for.

Answer It is of a great help, but in my company we have *excellent* people who are successful consultants and are not (and will never be) Fellows of the SOA. Of course this is rare, but it shows that it can happen. Personally, I hope I can make it to the end of the exams system and become a Fellow. But I sincerely think that I wouldn't be any worse off as a consultant in asset management if I were not a Fellow. I know people who are Fellows, but are not very good consultants because they are missing other essential skills.

Answer Exams bring recognition. Only the title is important. Not having the title makes things a bit more difficult, but not impossible.

Answer If you want to advance in a company, you need to be at least an Associate. I see many successful actuaries who are in charge of many clients and have stopped writing exams after their Associateship. But it takes them more time than it does a Fellow to get to the same point. It depends a lot on what your goal is and what other skills you have. A Fellow without interpersonal skills would probably lose clients.

Answer Consultants emphasize a lot more the importance of exams, since you are not allowed to sign anything for any clients until you become a Fellow. In insurance, I have met many people who chose to stop writing exams, and who have great jobs, and great positions in the company. Therefore, it depends on the goals of the student.

Answer I think it is directly correlated with how fast you can move up in a company such as a pension consulting firm. Showing dedication towards passing exams combined with gaining experience on the job can make for faster promotions.

Answer Usually, Fellow actuaries have jobs that carry more responsibility. For example, in an insurance company, they are often head of a department. As such, they coordinate the work of a team. In a consulting office, they are senior consultants. This means that they manage projects, meet clients and do the reports. When you are not a senior consultant, you are the one who does the evaluations, a more repetitive task.

Answer Not as important in group insurance benefits since most of our work does not require the signature of a Fellow.

Answer I do not believe that it is absolutely necessary to finish the actuarial examinations. In some of the companies where I have worked, I have met many actuarial professionals who stopped at Courses 4 [2002], 5 [2002], or 6 [2002] for reasons of their own. This has not stopped them from getting promotions or acquiring more knowledge. However, one cannot say that completing one's actuarial exams has no effect. From my experience and from what I have seen, successfully completing exams fast-forwarded more than one person's advancement.

Answer Canada: in life insurance company or pension consulting: can't move past analyst level without finishing exams. In casualty consulting: ACAS (or close) with experience can be a consultant, but can't sign reserves. In small/large casualty insurance companies: can't move into actuarial management, but may be able to move into non-actuarial management, e.g., underwriting.

Answer You need to get them all to become an actuary. The industry demands that. Careers can be very valuable to a company—in some cases they do the same work as actuaries, but are not allowed to officially sign anything. I feel that work experience is more valuable than exams. In the United States, individuals who have their ASA can call themselves *Actuaries*. There are many state commissioners who only have ASAs, but have a wealth of experience and are just as capable as their FSA counterparts. But companies don't hire people to become Career ASAs—they expect them to finish their exams.

Answer No exams: must concentrate on other skills such as communication, marketing, etc. Up to four exams: proven mathematics background. ASA: may be sufficient for many actuaries. FSA: allows you to use the designation fully; opens all the doors.

Answer I chose to stop at the ASA level and I'm really happy about it. I'm pursuing a career as a human resource specialist in employee benefits, and although I do not have the title to sign any official paper, I feel that I have enough experience and knowledge to smartly challenge our consultants, and not just sit back and listen to what they have to say.

Answer There is a big gap between an FSA and a Career ASA. If you are progressing steadily toward FSA, then any difference in pay, related to number of exams, will even out by the time you are fully qualified.

Answer It really depends on the organizations that you work for. As mentioned in a previous question, in my current organization, there are many actuaries that have not finished all actuarial exams that have excellent positions in other departments such as claims, underwriting, information technology, and finance (all Senior Vice-Presidents or Vice-Presidents). Also, within the actuarial department, there is one business unit that is run by actuaries without many exams. In other companies,you wouldn't see that.

Answer In consulting, examinations are very important if one wants to be a consulting actuary entitled to sign valuations. In large companies, it is harder to "move up the ladder" if you aren't a fully qualified actuary. It seems that in the past, ASAs were considered for senior positions. Based on what I am seeing in my company nowadays, that isn't the case. Qualifying is very important. Colleagues that choose to not complete exams are eventually placed in more technical positions, with lesser responsibility.

Answer At the end of the day, the number of actuarial examinations passed shouldn't matter, since one can learn the same thing on the job. However, in most cases, it does matter. One prior company I have worked with rewarded good work rather than exams. There, only two of the six managers were Fellows, the others having stopped writing exams. In most other places, having a Fellowship will tend to open doors. In some instances, Fellowship is an absolute requirement because of regulations (signing of valuation report, rate filing with regulatory bodies). In Canada, where the Associateship is not recognized, you almost need your Fellowship to become a consultant. Generally, Fellows will tend to move up the corporate ladder faster. Fellowship tends to indicate dedication, creative thinking, hard working qualities. However, there are always exceptions, and a Fellowship is generally not an indication of good management skills (besides time management). Companies would be well advised to look outside the box for promotion. All this being said, I do believe that passing examinations is still very important for an actuarial career.

Answer The number of examinations passed usually increase an actuary's responsibilities in everyday duties. More examinations represent more knowledge as well as commitment to the profession and to one's career. An example is someone whose title changed to Director of Actuarial Services when Associate status, as well as sufficient experience, were reached. The new position encompasses managing the projects of a team of actuaries and reporting to the Actuarial Vice-President.

Answer I believe that the number of actuarial examinations passed is important. The degree of importance varies from one company to the other. A person who has a good success rate with the exams is viewed as a very serious person. All the actuaries appreciate the level of difficulty, discipline, and hard work necessary to pass those exams. Therefore, a student who consistently writes them and passes them is viewed as a disciplined person and a hard worker. Depending on the type of work, not becoming a Fellow can stop one's progress within an organization. I work for a consulting firm. Some clients specifically ask for a Fellow to perform certain assignments. I work in the life valuation area. The appointed actuary needs to be a Fellow. Therefore, the fact of not becoming a Fellow has a great impact on one's career path because this person will never be able to become an appointed actuary. Moreover, at our firm, certain job levels can only be reached by Fellows. I have given examples on how being a Fellow is important. However, there are many companies where this is not as big an issue. I know a lot of actuarial people who did not become Fellows but still have a great career. Not every job requires a Fellow. However, on a personal level, I do believe that becoming a Fellow is important. I chose to study in actuarial mathematics so I could become an actuary. Never becoming an actuary would have meant not completely reaching my goal.

Answer CAS only: Comparing missing one exam with having passed all exams, the difference is like night and day. Here are a few reasons to pass all exams:

- *Promotion.* More likely to be promoted.
- *Opportunities.* Numerous new opportunities (both internal and external). Generally, some exams, such as pricing and reserving, are key to the work done in P/C. Therefore, having passed them (or one of them)—although nothing is automatic—may lead to more responsibilities, a promotion, getting staff, etc.
- *Salary.* Usually, the salary and the title are a function of experience and the number of exams passed. Companies usually apply varying weights to the two components. However, one thing is for sure, more exams usually means a faster progression through the ranks.
- *Pitfall.* There is one pitfall to having passed a lot of exams, however. If you have no or very little experience, and have passed a lot of exams, your "employment cost" may be too high. Therefore, you may end up being offered a lower salary than what you should really get, or some employers may decide not to offer you a junior position because it would be too costly considering your lack of experience. I consider that four exams—up to a maximum of five—is not too many for someone without any experience.

◆◆◆

1.10 **Associates and Fellows**

Q What can a Fellow do in your company that an Associate is not able to do?

Answer Not much in asset consulting, maybe some asset/liabilities studies.

Answer Signs reports. Is a client manager. Grows faster in the company.

Answer Not a lot, if the Associate has good complementary skills.

Answer A CAS Fellow can sign the actuarial reserves and the financial projections that all registered companies are required [to complete] by law. An Associate will know, and will perform those calculations, but will not be able to sign.

Answer Besides signing valuation reports, I'm not sure.

Answer We are required to be a Fellow to sign actuarial valuations.

Answer SOA and CAS Fellows have more credibility and are given more responsibilities than Associates. Therefore, Fellows will often be given the final say in decision-making situations and are asked to give their judgment on a particular project. They are more trusted. As if they could *do* more than Associates. However, I believe that the *doing* is strictly correlated with the number of years of experience rather than the number of exams passed.

Answer Sign reserves and rate filings—manage other Associates or Fellows. Otherwise, no difference in the work or opportunities.

Answer A lot—an ASA can't provide an official opinion on anything. Also, ASAs are thought of as FSAs in training. So they aren't given the same responsibilities that an FSA has. Career ASAs are often put into other (non-actuarial) roles in the company.

Answer Sign actuarial valuation reports!

Answer In Canada, an Associate of the Casualty Actuarial Society does not have any more legal authority or signing power than somebody with only a few exams. It is only once actuaries become Fellows (FCAS) and then receive their Fellowship of the Canadian Institute of Actuaries (FCIA) that they can start to legally sign documents. Examples are: Ontario automobile rate filings, year-end reserve valuations, and DCATs [dynamic capital adequacy testing].

Answer My department could not afford an FSA!

Answer Sign the actuarial valuations of a pension plan! This is the main job of an actuary!

Answer Sign business unit valuation reports. Be the Appointed Actuary for a subsidiary.

Answer I guess only what they can't do because of legal constraints (e.g., sign an Ontario rate filing or the Appointed Actuary's report).

Answer FSAs can sign off on actuarial valuations as a main signature, ASAs can only cosign on certain projects. FSAs appear to be given more senior roles and greater responsibility.

Answer In my former company, the only thing an Associate could not do was sign the year-end reports or rate filings filed with regulators. This is only because regulations require Fellows to do that. But there is nothing that says they cannot do the actual work, with the work being reviewed by the Fellow who will sign the documents.

Answer Fellows can sign actuarial valuations. They can be Appointed Actuaries. Some clients specifically ask for Fellows. Therefore, these assignments are not available to non-Fellows. The higher job levels can only be reached by Fellows.

Answer (CAS only) Tangibly: sign actuary's reports, sign rate-filings for the Financial Services Commission of Ontario, call themselves *Actuaries*, act as proctors for actuarial exams. Intangibly: receive greater trust from non-actuaries because of "perceived superiority." In actual fact, it makes no difference.

Answer Sign reports.

◆◆◆

1.11 Going for a Master's

In countries where the education of actuaries is university-based, it is often customary to pursue graduate studies at the Master's level. Having a Master's degree is in some sense equivalent to having achieved Fellowship status in a professional society. A Master's degree is required, for example, to become an Appointed Actuary in a country like Denmark. Moreover, university diplomas in many countries of continental Europe such as Germany are equivalent to Master's degrees in the United Kingdom and North America. However, some countries are beginning to introduce Bachelor's programs in actuarial science. In Austria, it is now possible to obtain a Bachelor's degree in actuarial science.

In countries where the accreditation of actuaries is based on professionally set examinations, the need for higher degrees is not obvious. In Canada and the United States, actuaries have to spend many years studying for their professional examinations. As a result, there is little incentive for acquiring a higher degree after that. Here is what some of respondents to the survey had to say:

Q **Discuss the value of graduate studies in actuarial science, accountancy, finance, economics, MBA, and so on, for an actuarial career.**

Answer My actuarial background has enabled me to leverage my MBA degree into a very valuable career as a Management Consultant. My post-MBA career, however, is not related to the actuarial discipline.

Answer I don't believe that it is of great importance. Maybe an MBA is helpful, but in my case, writing CFA [Chartered Financial Analyst] exams was of greater value for a career.

Answer Very limited value. Experience makes a good consultant, not study.

Answer Studying in actuarial science helps a lot for the exams. I also believe that it is easier to start. We have enough material to learn when we start, if we at least know all the actuarial science material, I believe the progression will be faster. But it is important to take courses in other fields to complete our knowledge.

Answer I do not know anyone with a graduate degree.

Answer I think that the SOA and CAS exams can be much more helpful in an actuarial career than any graduate studies. The only exception may be for actuaries who work in the finance field.

Answer In the actuarial field, SOA and CAS examinations have a greater value than graduate studies in actuarial science or other. However, they are not overlooked, and I consider it as a sign of a person's interest in learning and furthering their education, a fact that can only be applauded. A deeper understanding of actuarial science, finance, economics, etc., cannot hurt a future actuary but is not as essential as the SOA and CAS examinations.

Answer Good if planning a more research oriented career. I think there are many needs to fill.

Answer For me, the most useful tool would be project management. Other than that, being a recognized actuary (even ASA) is enough, most of the time, to add serious weight to the advice you give.

Answer I don't know yet. I don't think that I will pursue a Master's degree in actuarial science; I have yet to encounter a person in the workforce who has done so. I have not closed the door to the other degrees mentioned.

Answer Very little in business.

Answer In my opinion, graduate studies have very little value in promoting a successful actuarial career. It is not worth the time.

◆◆◆

1.12 | Career Alternatives

What if you have spent many hours studying to become an actuary and at some point you simply say to yourself that it is time for a change? Is there anything else you can do with all of this specialized training and knowledge? Here are some alternative career options:

Q What are the alternative professional options for actuaries who decide not to pursue an actuarial career? Please give examples.

Answer Management consulting (strategy or financial consulting with any of the major consulting firms, requires adaptability, ability to work well in teams, demands strong stamina to work long hours while traveling sometimes extensively), investment banking (demands strong interest in finance and ability to work long hours under pressure), asset securitization, risk management. An actuarial background and training provides a tremendously strong springboard to opportunities in a variety of non-actuarial disciplines.

Answer Investment managers, client relations for banks, managers, consulting firms, teacher, stockbroker, team leader.

Answer Mathematics teacher, investment consultant.

Answer Investment banker, portfolio manager, accountant, statistician, professor, researcher, economist.

Answer There are many fields available: teaching, research, finance, programming (for actuarial companies).

Answer I think an actuarial degree keeps many doors open as it shows an ability to understand complex concepts and apply theory to practical business problems. I know actuaries who have become directors of human resources for large companies. This position would involve the company's compensation practice. Also, some actuaries decide to enter the finance field.

Answer   Finance (asset management), human resource advisor, teacher.

Answer   Graduating with a BSc or BA in actuarial science is a great basis to later on continue to an MBA. Any profession in finance and management can be pursued. As well, becoming a CFA [Chartered Financial Analyst] or chartered accountant are possibilities. Of course, the option of working for an insurance company is always there.

Answer   There are many professions that use the mathematical skills, technical skills, software skills, and problem-solving skills that are supposed to be developed in the actuarial education process. Certified financial analysts, statistical modeling specialists, forecasting modeling specialists, computer programmers, etc. Generally in investment or IS-type employment. Could also sell insurance with an actuarial background.

Answer   Financial analyst with investment managers (requires the CFA [Chartered Financial Analyst] designation), director of employee benefits with a private firm, IT [information technology] specialist (requires strong software skills), banking specialist.

Answer   Banking, finance.

Answer   Interesting question. One would be to pursue the Chartered Financial Analyst designation. There are three exams and you can work in more investment related fields and skip the mathematics portion. Another would be to go to graduate school in order to teach or to get a degree in financial engineering. The field is opening a lot to nontraditional positions. I'm not that familiar with them.

Answer   Produce special quotes; be more systems-oriented (if you enjoy programming); become a manager in an insurance company where business skills are very important, doubled with actuarial knowledge.

Answer   Teaching, risk management (derivatives, hedging programs, portfolio management), teaching mathematics at the high school level.

Answer   I believe that there are many opportunities. Within my current organization, there are many actuaries outside the actuarial department: claims, underwriting, investment, and IT [information technology] (all vice-presidents). Outside the insurance world, you could probably become a financial advisor, a management consultant, or a risk management consultant.

Answer Here is a list of professions some of my friends chose after deciding not pursue an actuarial career: banking, investment banking, financial advising, brokerage services, pension administration, programming specialists, investment consulting.

Answer Teacher, statistician, finance-related profession, programming, etc.

Answer Engineering, consulting, finance/investments, banking, statistician, math teacher/professor.

Answer Programmer, teacher, career in research, CFA [Chartered Financial Analyst]. There are definitely many career possibilities for someone who has passed actuarial exams (especially if all of the exams have been passed).

Answer Teaching, working in finance.

Answer Retraining would be necessary if an actuary expects to be compensated as highly, but there is no limit. There are a number of ancillary positions within the insurance/pension industry to which actuaries could easily apply their skills. Sales/marketing, software development, accounting, business planning. They could also readily move into another part of financial services and become stock analysts or financial engineers.

Answer Computer programming, marketing.

Answer Teaching at the college and university level. Investment: I know a few people, they wrote their CFA [Chartered Financial Analyst] exams and now work in marketing or servicing departments of investment management firms.

◆◆◆

1.13 | Actuaries around the World

As is the case with most formally structured professions such as medicine, engineering, accounting, and others, the international employment of actuaries involves two critical elements: a recognition of academic qualifications attained in another country, and a license to practice.

In Europe and Latin America, actuaries have tended to qualify by completing a course of actuarial study, usually up to the Master's level, at universities accredited by actuarial societies or governments, and by meeting certain professional requirements.

In Great Britain and Commonwealth countries, the Institute and Faculty of Actuaries has defined an actuarial syllabus and sets of examinations based on this syllabus. Students in many parts of the world take these examinations to become actuaries in their countries. Certain university courses at designated universities can be credited towards this process.

In the United States, the Society of Actuaries and the Casualty Actuaries Society have defined a syllabus and sets of examinations that must be taken to become an actuary. No university program or courses are credited toward these examinations. Most American and Canadian actuaries and many others around the world become actuaries by passing these examinations.

As a result of changes in the world economy, several countries with a university-based accreditation system are looking toward instituting nationally administered examination systems, whereas countries with professionally based accreditation systems are considering the idea of granting exemptions for some courses taken at university. The dynamics involved in these deliberations are outlined in greater detail in Brown's paper on the globalization of actuarial education (see [6]).

The process of becoming an actuary is far from uniform. When contrasting the North American actuarial education with that of Europe, for example, one working actuary explains that "in general, the European programs are more like graduate programs or short seminar-and-test programs than the lengthy study-while-you-work exam system of the United States. As a result, the actuarial training tends to be more theoretical than hands-on, and the resulting actuaries are very strong in statistics, modeling, etc. American actuaries tend to have more real-life experience. Some of this focus may be due to the available data (at least from a P/C perspective). Regulatory requirements force insurers in the United States to collect much detailed data, (usually) suitable for analysis. European insurance data is not as voluminous, so there is more emphasis on theory, modeling, and the like. Purely anecdotally, the French actuarial education may be the most theoretical, while the UK exam system is the one that most resembles that of the United States. However, the United Kingdom does not make the distinction between Life and P/C (Non-Life, as they call it here) made in the United States (SOA versus CAS), and I believe many of the European actuarial societies also take that approach."

Nevertheless, it is actually fairly easy for actuaries educated in North America and the United Kingdom to work in other countries. The following examples will give you some idea of different national accreditation systems and of the portability of the acquired qualifications between certain countries.

Argentina

The education of actuaries in Argentina is university-based. It is necessary to obtain a degree of *Actuario* at a local recognized university in Argentina, like the University of Buenos Aires, which follows the syllabus recommended by the International Actuarial Association (see Chapter 2).

Nowadays the University of Buenos Aires issues diplomas following two orientations: *Actuario-Administracion* and *Actuario-Economa*, each of which provides an actuary with a license to practice. The degrees differ only in some courses on administration and economics.

To become accredited actuaries, graduates must register their diplomas with the *Consejo Profesional de Ciencias Economicas* of the State where they would like to practice as independent consultants or be employed in a position that requires an actuarial degree, such as the *Consejo Profesional de Ciencias Economicas de la Ciudad Autonoma de Buenos Aires*, for example.

Each State in Argentina has a public professional council called the *Consejo Profesional de Ciencias Economicas*, created by law, which controls the independent activity of accountants, actuaries, administrators, and economists. These councils are responsible for maintaining professional standards. In order to be able to issue independent reports and formal advice, actuaries must usually be registered with the Consejo of the State in which they work. Registration requires an actuarial diploma issued by a recognized Argentine university (private or public) or a diploma from a foreign university recognized by a public university with a full actuarial program, such as the University of Buenos Aires.

Australia

Admission as a Fellow of the Institute of Actuaries of Australia (FIAA) is granted once all five parts of the Institute of Actuaries of Australia's (IAAust) education program are successfully completed: (1) Part I—Technical Subjects. (2) Part II—The Actuarial Control Cycle. (3) Part III—Specialist Subjects. (4) The Practical Experience Requirement. (5) Professionalism Course.

Part I is made up of nine subjects including statistical modeling, financial mathematics, stochastic modeling, survival models, actuarial mathematics, economics, finance, financial reporting, and financial economics. All nine subjects must be completed.

Accredited undergraduate actuarial programs and non-award courses are offered by Macquarie University, Sydney, the University of Melbourne, the Australian National University (ANU) in Canberra, and the University of New South Wales (UNSW) in Sydney. Alternatively, these subjects can be studied by correspondence through the Institute of Actuaries (London).

Part II of actuarial education is the actuarial control cycle, which is an innovative means for learning how to apply actuarial skills to business situations across a wide range of traditional and nontraditional practice areas. Developed by the IAAust, this course is taught by four universities in Australia (as mentioned above). A strong and rigorous policy framework for accreditation of the university courses is in place, so that the IAAust maintains quality control of the teaching and assessment of the courses. After completing Parts I and II, members achieve Associateship of the IAAust (AIAA).

Part III consists of specialist subjects, of which students must complete two, in life insurance, general insurance, superannuation and planned savings, finance, and investment management. These yearlong courses are developed and managed by the IAAust and are offered by distance education.

Students must complete 45 full-time working weeks of relevant work experience after having completed Part II. Activities that qualify as relevant experience would include work that makes use of economic, financial, and statistical principles to solve practical problems; and work that deals with the financial implications of uncertain events.

The professionalism course is a highly participative three-day residential course conducted by the IAAust. It aims to facilitate knowledge of the obligations, risks, and legal responsibilities of being a member of the actuarial profession.

The IAAust has concluded a number of bilateral agreements for mutual recognition of Fellows with the Institute and Faculty of Actuaries (UK), the Society of Actuaries, the Canadian Institute of Actuaries, and the Society of Actuaries of Ireland.

These agreements enable actuaries to practice professionally in other territories subject to meeting the requirements of the local actuarial association. Each agreement is predicated on equivalent educational and professional conduct standards. In addition, a period of professional practice and residency within Australia is required prior to overseas actuaries being eligible to attain full Fellowship status of the IAAust.

In order to become Associates, candidates must pass Parts I and II of the actuarial examination curriculum. Graduates from Australian and New Zealand universities must either be math majors or Honors graduates in another discipline, provided they have attained a sufficiently high standing in mathematics as part of their degree program. The Australian National University is accredited by the Institute of Actuaries of Australia to provide students with exemptions from certain examinations of the Institute, provided the students obtain sufficiently high grades in designated courses.

Austria

The Austrian equivalent of a Fellow of the Society of Actuaries is that of an *Anerkannter Aktuar*, a regular member of the Actuarial Association of Austria. To become a member of the Society, a candidate must have obtained a university degree in actuarial science and have three years of professional actuarial experience. The Technical University of Vienna is the only Austrian university offering novel degree programs in actuarial science following the North American Bachelor's and Master's degree structure. Austria is a member of the *Groupe Consultatif* and Austrian actuaries can take advantage of the European reciprocity agreements coordinated by the Groupe. Foreign-trained actuaries can become members of the Association if their academic training and professional experience meets the requirement of the Association. The main function of the Austrian Association of Actuaries is to promote the education and training of its members, to represent actuaries both nationally and internationally, and to establish guidelines and rules for good actuarial practice in Austria.

Belgium

The Belgian equivalent of a Fellow of the Society of Actuaries is that of a full member of the Belgian Society of Actuaries (KVBA-ARAB). How do you become such a member? Candidates must first obtain a Bachelor's and a Master's degree in mathematics, economics, civil engineering, or physics (which takes four to five years at university). They can then be admitted to an actuarial program recognized by the professional organization (run by ULB in Brussels, UCL in Louvain-la-Neuve, KULeuven in Leuven, or VUB in Brussels). The students need at least two years to complete the program. On the basis of this academic training they can become junior members of the Belgian Society of Actuaries. They then need three more years to become full members. During that time they must follow certain courses run by the professional association on professionalism, code of conduct, and other topics. They are expected to follow similar activities during their career, but this is not yet compulsory.

The academic programs will probably be modified in a few years' time because of new European guidelines concerning actuarial studies. As a result, actuarial studies in Belgium may become two years at Master's level, and students with a Bachelor's degree (three years of university study) will be admitted to the program. It will then take five years at university to become an actuary instead of the current six to seven years.

The *Groupe Consultatif* representing the national professional associations of the Free European Exchange zone (the so-called "Espace Economique Europeen," which is larger than the EU) has set up a mutual recognition agreement (see Appendix C). A crucial step is the creation of a European program for actuarial studies. But this is for European actuaries (in the broad sense). For overseas actuaries, the rule is that they must apply to the Education Committee of the Belgian Society of Actuaries for recognition of the equivalence of their credentials to the Belgian requirements. Their level of training is then compared to the Belgian one. If comparable, foreign actuaries are admitted, provided they will work in Belgium.

Brazil

The education of actuaries in Brazil is university-based and is offered only at the undergraduate level. The focal points are Rio de Janeiro and São Paulo. The profession is loosely organized through the Instituto Brasileiro de Actuária. The institute represented Brazil at the first international professional meeting of leaders of the actuarial profession and actuarial education in Latin America, held in Buenos Aires, Argentina, in 2002. Brazil has no actuarial Fellowship system and certain aspects of life insurance and reinsurance are government run.

Denmark

The Danish Society of Actuaries was established in 1901. It works closely with the *Groupe Consultatif* and its main objective is the advancement of actuarial science and to promote the interests of the actuarial profession in Denmark. The society participates in the government supervision of financial institutions and is represented on the Ministry of Economic Affairs. It participates in all hearings on actuarial concerns and is often represented on government-appointed committees. Actuaries in Denmark are usually divided into non–life insurance actuaries and life insurance actuaries. All life insurance companies and pension funds must employ an Appointed Actuary approved by the Danish Financial Supervisory Board. Actuarial education in Denmark is university-based. Most actuaries in Denmark have a Master's degree in actuarial science from the University of Copenhagen. To become an Appointed Actuary you must have a Master's degree (not necessarily in actuarial science) and, in addition, at least three to five years of insurance experience. Theory and practice in actuarial science in Denmark are closely linked since many professors teaching at the Laboratory of Actuarial Mathematics at the University of Copenhagen are sometimes also employed by insurance companies and are active members of the Danish Actuarial Society.

Finland

The actuarial profession in Finland is organized through the Actuarial Society of Finland, which has approximately three hundred members. About one third of them are fully certified actuaries. However, the government formally controls the actuarial education and actuarial accreditation.

The Ministry of Social Affairs and Health of Finland nominates an Actuarial Examination Board, which administers relevant examinations and controls the syllabus and the qualification standards. The Ministry works closely with the Actuarial Society of Finland in the sense that the members of the Examination Board, for example, are usually also members of the Actuarial Society. However, the Insurance Supervisory Authority has additional resources for developing actuarial education and research, upon which the Board also draws.

Admission to the Fellowship of the Actuarial Society of Finland is granted on successful completion of a relevant university degree, the completion of actuarial foundation courses, the passing of additional examinations dealing with actuarial applications, the writing of a thesis, and the completion of at least one year of practical actuarial work.

Universities offer the foundation courses, whereas the Examination Board prepares the actuarial application examinations. Foundation courses cover risk mathematics, survival models, financial mathematics, and basic life insurance. Many of the examinations are written while the candidates are fully employed, so that it usually takes several years before they are able to qualify for Fellowship.

To enter the actuarial profession, graduates must have a Master's degree with a major in mathematics or cognate discipline, provided that a sufficiently high standard in mathematics has been demonstrated. Courses in probability, statistics, and stochastic processes are particularly relevant. Only some universities offer the actuarial foundation courses. The University of Helsinki, on the other hand, also has an M.Sc. program in mathematics, with specialization in actuarial studies.

In order to implement the Groupe Consultatif Core Syllabus, changes have been made in the education system. Courses on financial economics and investment mathematics have become mandatory and a course in economics has been added to the syllabus.

The examinations in actuarial applications include four general examinations and an individualized self-study test.

At the general level, the subjects of the tests are insurance legislation, insurance accounting, and applied insurance mathematics. These examinations are country-specific since they are based on Finnish insurance legislations. The applied insurance mathematics examination covers actuarial modelling, practical risk theory, solvency issues, and investments.

At the specialized level, candidates select one of the following subjects: life insurance, mandatory pensions, or general insurance. Along with general principles and practice, Finnish conditions are emphasized. This holds, in particular, for pensions because of Finland's unique mandatory pension system.

The Examination Board supervises the thesis. Starting in 2005, two options became available: First, candidates are able to write a brief analysis of an issue of actuarial concern. The purpose of this type of thesis is to demonstrate the ability to present ideas and arguments. The second option is to write a research paper on a practical topic. Many candidates have a great deal of work experience before attaining full professional status. The aim of this type of thesis is therefore to encourage the development of that experience and foster innovations in actuarial science. The Foundation for Promotion of the Actuarial Profession actually encourages this type of work by providing financial support.

Finally, candidates must have completed at least one year's practical experience in an insurance company or have done equivalent work. Experience may count as equivalent if it consists of practical applications of actuarial methods, under the supervision of a Fellow of the Society of Actuaries.

Continuing professional development is not mandatory. The Society offers voluntary seminars and courses on topical issues when legislation is changed or the environment is changed otherwise. Most actuaries attend these activities. Another popular form of professional development is participation in actuarial conferences.

France

The education of actuaries in France is university-based. Three universities offer degree programs in actuarial science: Brest, Lyon, and Strasbourg. According to Morgan (see [16]), "The profession is still underdeveloped compared to the United Kingdom, and France is the only European country where actuaries are not a legally recognized profession." Morgan points out that "as in many European countries, the actuarial profession has been more academic and less practical than that in the United Kingdom, but this is changing as elements of accounting, law, and tax have been added to the course of study. These days, actuaries work in banks and consultancies as well as in insurance companies. In insurance their role is widening to include marketing and communication as well as just technical matters such as ALM [asset and liability management], and embedded values are starting to become more widespread."

Germany

Germany has its own version of professional accreditation. In order to qualify for membership in the *Deutsche Aktuarvereinigung* (Actuarial Association of Germany), candidates must pass examinations testing their general and specific competence in actuarial science. The *Deutsche Aktuarvereinigung* has joined forces with the *Deutsche Gesellschaft für Versicherungsmathematik* (German Society for Insurance Mathematics) and the *Institut der Versicherungsmathematischen Sachverständigen* (Institute of Experts in Insurance Mathematics) and founded the *Deutsche Aktuar-Akademie* (DAA) (German Actuarial Academy), which provides basic and advanced training for actuaries. The DAA holds seminars and workshops for the courses in which actuarial candidates are examined.

The German accreditation system consists of three levels of examinations, each consisting of several courses. Each level is considered to require one year of preparation. Level 1 consists of three examinations and one compulsory course in data processing. The subjects examined include mathematics of the life insurance, mathematics of finance, and other elementary actuarial topics. Level 2 consists of two examinations, chosen from four topic areas: P/C, pensions and stochastic methods, real estate, and health. Level 3 consists of a compulsory seminar and examination in one of the following specialties: life insurance, P/C, pensions, applications of stochastic methods, health, and finance. Several German universities offer degree programs in actuarial science. Among them are the universities of Ulm and Göttingen.

Hong Kong

The Actuarial Society of Hong Kong does not conduct its own set of actuarial examinations at the moment. It relies on the exam systems of other established overseas actuarial bodies. Typically, to be admitted as a Fellow of the Actuarial Society of Hong Kong, the member must be a Fellow of one of the actuarial bodies of Australia, Canada, the United Kingdom, or the United States, although there is an increasing number from other countries, especially from Europe. Under the Hong Kong government's insurance company (qualification of actuaries) regulations, the qualifications for appointment as an Appointed Actuary are: Fellow of the Institute and Faculty of Actuaries in the United Kingdom, Fellow of the Institute of Actuaries of Australia, or Fellow of the Society of Actuaries of the United States of America.

Until recently, students who wished to pursue an actuarial science degree had to travel overseas. Now three universities—the University of Hong Kong, the Chinese University of Hong Kong, and the Hong Kong Polytechnic University—offer a range of actuarial subjects.

India

The actuarial education in India is profession-based. The Actuarial Society of India offers a series of examinations that must be passed to qualify as an actuary. The Society was established in 1944 to provide a central organization for actuaries in order to raise the standards of competence and level of recognition of the actuarial profession. The structure of the Society and its examination syllabus are comparable to that of the Institute of Actuaries of the United Kingdom. As in other countries, actuaries in India are involved in insurance, pensions, investment, financial planning, and management. According to the Society, actuaries have "an unlimited scope in countries outside India where the necessary infrastructure already exists to absorb them in suitable avenues like life and general insurance, operations research, statistics, investment, demography, etc. The remuneration offered is very lucrative and the job satisfaction is tremendous."

Ireland

The equivalent of a Fellow of the Society of Actuaries is a Fellow of the Society of Actuaries in Ireland (FSAI). Most Fellows qualify through the Institute and Faculty of Actuaries in the United Kingdom. Under the constitution of the Society of Actuaries in Ireland, all Fellows must be Fellows of the Institute and Faculty of Actuaries of the United Kingdom, or they can join the Society through mutual recognition agreements if they are Fellows of the *Groupe Consultatif*, the Australian Institute, the Society of Actuaries, or the Canadian Institute of Actuaries. A foreign actuary can join the Society via a mutual recognition agreement with one of the aforementioned bodies.

Israel

The education of actuaries in Israel is concentrated in universities. In addition to courses in actuarial science available at the Hebrew University in Jerusalem and at Tel Aviv Univesity, the University of Haifa maintains an active research center in actuarial science, offering a Master's degree. The Israel Association of Actuaries is the professional body for actuaries in Israel and is a full member of the International Actuarial Association. One of its functions is to enhance the practical knowledge of graduates of the academic courses in Israel and abroad and examine these candidates for Fellowship. Individuals with actuarial training can also become qualified members of the Society of Actuaries of the United States by writing examinations at the permanent SOA examination center in Ramat Gan.

According to the historical account of the evolution of the actuarial profession in Israel [see http://hevra.haifa.ac.il/en], "the Israeli industry's approach to financial risk has consisted of adapting foreign solutions to better reflect Israeli reality and its needs. Thus, the mortality tables in use in Israel today come from England and are subject to an adjustment. However, this adjustment has no scientific justification or basis; at best, it represents the intuition of insurance company actuaries or, alternatively, it is a manifestation of the interests of such companies, a possibility that has drawn ample criticism. In the Western world, actuarial centers work to gather and analyze mortality data to provide the mortality tables necessary for performing precise calculations. In Israel, this step has yet to be taken. The Actuarial Research Center aims to close this gap."

Italy

Actuarial life in Italy is coordinated through the *Istituto Italiano degli Attuari* and by *Ordine Nazionale degli Attuari*.

The Institute is a member of the *Groupe Consultatif* and therefore has reciprocal agreements with the member countries of that group.

The Italian actuarial associations maintain a permanent professional development program through the *Corsi di Formazione Attuariale Permanente Program*, allowing its members to keep up-to-date with changes in actuarial practice resulting from globalization and European integration. The actuarial education in Italy is university-based and the title of fully qualified actuary is obtained through a state examination. To act as a consultant an actuary must be enrolled in the National Register (*Albo Nazionale*), established by law in 1942.

The state examination program is under review to be consistent with the recent reform of the university system.

Japan

The actuarial education is profession-based. The Institute of Actuaries of Japan offers actuarial courses that enable applicants to acquire basic knowledge and to prepare for qualification examinations. Actuarial courses are divided into two categories, basic and advanced courses. The basic courses are intended for students of the Institute, while advanced courses are aimed at persons who have completed the basic subjects.

To become an Associate member of the Institute, candidates must pass examinations in the following five basic courses:

1. Probability and statistics.
2. Basic principles and applications of life insurance mathematics.
3. Basic principles and applications of non–life insurance mathematics.
4. Basic principles of pension mathematics and pension finance.
5. Basic principles of accounting, economics, and investment theory.

After passing these courses, candidates qualify for Associate membership in the Institute of Actuaries of Japan.

To become a Fellow of the Institute, Associates must pass two additional advanced courses: (LI1) Life insurance products and development and (LI2) Life insurance accounting, settlements of accounts, or (NLI1) Non–life insurance products and development and (NLI2) Non–life insurance accounting, settlements of accounts and asset management, or (P1) Tax qualified pension plan scheme and pension-related tax and accounting and (P2) Public pension system and employees' pension fund scheme. Fellowships are approved by the Board of Directors of the Institute. New Fellows are also strongly recommended to take a half-day professionalism course.

Several Japanese universities offer courses on actuarial mathematics and risk management, but there are no exemptions for qualification examinations.

Malaysia

The actuarial profession in Malaysia is represented by the Actuarial Society of Malaysia. The Society does not have its own accreditation process. Actuaries who are Fellows of the following institutions may be admitted as Fellows of the Society: the Institute and Faculty of Actuaries of the United Kingdom, the Society of Actuaries of America, the Canadian Institute of Actuaries, or the Institute of Actuaries of Australia. Admission to the Society must be approved by the Executive Committee of the Society. Qualified actuaries are allowed to practice in Malaysia if they reside in Malaysia or, in the opinion of the Executive Committee, are familiar with Malaysian conditions, and have paid the requisite admission and annual membership dues.

Fellows of the Society can become Appointed Actuaries of insurance companies by being approved by the regulatory authority in Malaysia (Bank Negra Malaysia). Appointed Actuaries must be residents of Malaysia and have at least one year of relevant work experience with a Malaysian insurer.

Mexico

The education of actuaries in Mexico is university-based. To be able to work as an actuary in Mexico, and to be allowed to use the designation "actuary," candidates must fulfill three requirements: (1) They must complete a four-year undergraduate program in actuarial science, which includes 480 hours of unpaid socially valuable work. The Mexican syllabus is close to that prescribed by the SOA. In fact, many students in Mexico are encouraged to write the SOA examinations. (2) They must write a relevant dissertation. (3) They must defend the dissertation before an examination committee.

Some universities accept graduate work in relevant academic programs and the passing of written and oral comprehensive examinations as dissertation equivalents. In order to be accredited as actuaries with signing privileges, graduates must have their university degrees approved by the Ministry of Education and obtain from the Ministry a *cédula profesional*. Certified actuaries are publicly sworn to uphold the code of ethics of the profession, but they are not required to become members of the Colegio or any other association of actuaries. A significant number of actuaries in Mexico work in nontraditional areas such as finance, government, planning, and information technology.

Mexico has two actuarial organizations: In 1962, the *Asociación Mexicana de Actuarios del Seguro de Vida* was formed. Its members tend to work in life insurance. In 1980, the association expanded its membership to include all actuaries and become the *Asociación Mexicana de Actuarios*. In addition, the profession established the College of Actuaries in 1867, the *Colegio de Actuarios de México,* which was transformed into the *Colegio Nacional de Actuarios* in 1982. Membership in the College is not required to function as an actuary.

Netherlands

The Dutch equivalent of Fellowship in the Society of Actuaries is a Fellowship in the *Actuarieel Genootschap*. Two roads lead to this Fellowship:

1. Successful completion of the actuarial program of the *Actuarieel Instituut*. This involves between eight and nine years of study.

2. The other option is to complete a Master's program in actuarial science at the University of Amsterdam. This involves between four and five years of study, together with a successful completion of the two-year post-Master's course of study administered by the *Actuarieel Instituut*.

The Netherlands has a reciprocal agreement with the *Groupe Consultatif*, which allows for recognition of the Fellowships of other member countries.

Norway

The education of actuaries in Norway is university-based. Since 1916, the University of Oslo has offered a degree program in actuarial science in insurance mathematics and statistics. In the 1950s and 1960s, a program of actuarial studies based on stochastic principles was established. Risk theory and non–life insurance were added to the syllabus in the 1970s. Now, actuarial students are required to complete a five-year Master's program in mathematical statistics with specialization in insurance mathematics. Students must also write a Master's thesis equivalent to one year of full-time work experience. The study of economics is no longer a required part of the program. The current thinking is to make the program more applied and introduce a business component. Since 1997, the University of Bergen also offers a degree program in actuarial science. Almost all actuaries in Norway belong to the Norwegian Actuarial Society.

Portugal

The Portuguese equivalent of a Fellow of the Society of Actuaries is an *Actuário Titular,* a full member of the Portuguese Institute of Actuaries. The education of actuaries is university-based. The Technical University of Lisbon offers courses in actuarial science. A graduate with a university degree in mathematics, economics, and management, together with the appropriate courses in actuarial science and three years of experience working as an actuary, can become a member. The candidate must prepare a report under the supervision of an accredited member of the Institute as part of the accreditation process. Foreign-trained actuaries must have their professional credentials recognized by the Institute to become members.

Singapore

According to its mission statement, "the Singapore Actuarial Society was formed in 1976. At that time, the profession was little known in Singapore and there were only a handful of qualified actuaries. The adoption of the new Constitution in July 1996 and the Code of Professional Conduct in November 1997 is the fruition of efforts made in the past two decades to promote the study of actuarial science and professional standards. The Society ultimately aims to be the recognized representative body of the actuarial profession in Singapore, having the final authority in setting professional standards.

The objectives of the Society are:

- to uphold the highest professional standards among members,
- to promote the study, discussion, publication and research into the application of economic, financial and statistical principles to practical problems, the actuarial, economic and allied aspects of life assurance, non-life insurance, employee retirement benefits, finance and investment with particular reference to Singapore and the ASEAN region
- to assist students in the course of their actuarial studies and
- to foster and encourage social relationship among the members."

The Society maintains a website at www.actuaries.org.sg.

South Africa

The life insurance industry was brought to South Africa by British settlers in 1820, and the first actuaries in South Africa were all Britons, employed by UK-based companies. As a result, South Africans who were interested in the actuarial profession were exposed to the UK Institute and Faculty of Actuaries. To this day, the vast majority of South Africans qualify via the UK actuarial education system. Applications for admission as a student member of one of the UK organizations are dealt with by the Admissions Committee of the Actuarial Society of South Africa.

Various South African universities offer undergraduate and postgraduate courses in Actuarial Science. Students taking these courses may obtain exemptions from most of the subjects required by the UK syllabus, depending on their university results. These exemptions make it possible for a student, after university, to have to write only one subject in the 300 series and the 400 series before qualifying as an actuary. A small number of South African students follow the courses prescribed by the U.S. Society of Actuaries and the Casualty Actuarial Society. Examinations of the UK and U.S. organizations are administered by the Actuarial Society of South Africa (ASSA) at centers in Cape Town, Durban, Johannesburg, and Windhoek (Namibia).

Since September 2003, students have the option of writing a Fellowship paper based on South African legislation and regulation, instead of the UK paper. The local-content paper is set in conjunction with the UK actuarial education authorities and results in the same qualification being awarded, i.e., either FFA or FIA.

Attendance at the ASSA professionalism course is required for all actuaries within a year of completing the exams of the Institute and Faculty of Actuaries. The course is recognized by the Institute and Faculty of Actuaries. The course structure is based closely on that used for the professionalism course run by the Institute and Faculty of Actuaries in the UK. To a large extent, the course material is identical to that used in the UK.

The aim of the ASSA Professionalism Course Committee was to construct a course that meets the Institute and Faculty of Actuaries requirements for recognition while providing sufficient local South African content to keep the course relevant and interesting for South African delegates.

In addition to the incorporation of case studies reflecting issues facing the profession in South Africa, the UK course material is supplemented by course material on ethics, as used by the Institute of Actuaries of Australia. The course material is further supplemented by locally developed material on legal liability and conflicts of interest.

ASSA's professionalism course is run twice a year. It is run over two days on a residential basis and comprises a total of around 12 hours of working time. The course is run by two lecturers (both experienced qualified actuaries) and a guest speaker (also an actuary). The guest speaker is chosen to talk about one of the wider fields (health care, general insurance, or investment work—depending on the number of delegates involved in each of these wider fields) and to illustrate issues of a professionalism nature facing actuaries in such a field. The course utilizes a variety of media, including lectures, group workshops, case studies, newspaper clippings, and video and audio material.

ASSA provides regular opportunities for its members to keep up with technical and professional developments. A *Continuous Professional Development Compliance Certificate* is mandatory for actuaries who are active in certain fields.

The Financial Services Board, as regulatory authority of the financial services sector in South Africa, approves actuaries as statutory actuaries of life offices and as valuators of retirement funds. In this process, the Board requires an applicant to submit a relevant practicing certificate, which is issued by ASSA.

South African law stipulates that valuations of life offices and retirement funds have to be performed by actuaries, but there is no formal, statutory actuarial involvement in other fields at present. Some progress has been made with regard to the formal involvement of actuaries in health care and short-term insurance.

The major fields of activity for actuaries practicing in South Africa are life offices (31%), employee benefits (28%), consultancy work (involving life insurance, health care and employee benefit work, but not employed by an insurer—21%), health care (9%), and investments (5%). The remaining 6% are involved in short-term insurance, academia, etc.

Spain

Since the establishment of the *Instituto de Actuarios Españoles*, the Spanish professional association, and still to this day, the only requirement to be admitted as a full member is to have the actuarial and financial sciences degree. This higher education degree, offered by several universities through their faculties of economics and business administration, is a full actuarial education program, which normally takes two years and consists of 150 credits (one credit involves ten effective lecture hours), where almost half of the credits must be in the following subjects: actuarial statistics (including topics on stochastic processes, survival models and, partially, risk theory), financial mathematics (also including topics on investment), actuarial mathematics (including risk theory and life and non–life insurance mathematics), accounting and financial reporting in insurance, banking and investment, insurance, banking and stock market regulations, and social security economics and techniques.

In deciding on the rest of the program or syllabus, each university has a significant degree of autonomy, but most of them expand the number of credits in actuarial mathematics, financial mathematics, statistics, accounting, and financial reporting, and then offer specialized courses in private pension plans, financial instruments and markets, taxation, solvency, reinsurance, insurance and financial marketing, computing, and so on.

Students who want to pursue such a program must have an undergraduate degree, usually in economics or business administration. It must include courses in mathematics, probability and statistics, economics, finance and accounting, and financial reporting.

In addition, the *Instituto de Actuarios Españoles* has implemented the changes required to meet the education requirements set by the European Union Recognition Agreement (*Groupe Consultatif* Core Syllabus). This program and the examinations involved will be set and run by the Spanish Institute of Actuaries through the "Actuarial Training School."

Sweden

In Sweden, the education of actuaries is based on a hybrid system. Membership in the *Svenska Aktuarieföreningen* is granted to individuals who have fulfilled appropriate academic requirements. (1) Candidates must have the necessary grades in basic mathematics and mathematical statistics. (2) Candidates must also have obtained an actuarial diploma, granted to a person who, in addition to what is required for step 1, has the necessary grades in actuarial subjects, including actuarial science, law, and economics; has written an approved actuarial paper; and has at least two years of experience working as an actuary. The required knowledge and experience may have been acquired in another country, and the *Svenska Aktuarieföreningen* is a member of the European cross-border mutual recognition agreement. License to act as an Appointed Actuary is granted by the *Finansinspektionen* supervisory board. The requirements are similar to those required for the actuarial diploma.

Switzerland

In Switzerland, the profession is made up, as everywhere else, of actuaries with university degrees who have passed special professional examinations and have special legislative powers assigned to them in their capacity as general insurance and life insurance actuaries. In the case of pension actuaries, the situation is even more structured. Pensions in Switzerland are subject to special laws and pension contributions are mandatory for all employers. This is an important difference between Switzerland and North America. As a result, pension actuaries must obtain additional qualifications to practice.

Swiss actuaries receive their qualification from the Swiss Association of Actuaries and hold the title "Actuary SAA." They are qualified in the sense of being full members of the IAA. If Fellows of the SOA want to work for pension funds in Switzerland, they must pass an appropriate special examination, in addition to being accredited by the **Swiss Association of Actuaries (SAA)**.

The academic and professional steps required for Fellows of the Society of Actuaries to be recognized as an Actuary SAA involve an evaluation of their credentials by an admissions commission of the SAA. They will decide whether there are deficiencies to be filled by extra examinations or whether all aspects of the

Swiss syllabus have been covered. In the first case, the candidate will be notified of the exams to pass, in the second case the admission is straightforward.

If Fellows of the Society of Actuaries want to be able to give statements to pension funds, they must submit their credentials to the examination commission of pension funds experts. Afterwards the procedure is the same as for the Actuary SAA, but there will, with great probability, be a deficiency in legal knowledge. Fellows of the SOA will therefore almost certainly have to pass at least the legal examination before receiving their state diploma.

Actuarial studies in actuarial mathematics in Switzerland involve completing a course of studies based on the Swiss syllabus. Certain Swiss universities are accredited by the SAA. An appropriate degree from these universities means that the academic requirements have been met. For students from other universities, the same procedure as for a Fellow of the Society of Actuaries is required: existing knowledge is evaluated and deficiencies have to be made up for by special examinations. In addition, actuaries must have three years of practical experience and have passed the examination colloquium.

For pension fund experts there are special admission examinations, preliminary examinations, and a comprehensive examination, all of which must be passed, with the possibility of taking into account former qualifications.

United Kingdom

The Institute of Actuaries in England and the Faculty of Actuaries in Scotland were the two professional associations which represented actuaries in the United Kingdom. In 2010, the two associations merged and created the Institute and Faculty of Actuaries. This association maintains and manages an examination and accreditation system not too different in content from that of the Society of Actuaries in the United States, with strong emphasis on professionalism and work experience. Here too the actuarial world is built on the four corners identified before: core knowledge, specialized knowledge, professional knowledge, and experiential knowledge.

The core technical knowledge comprises financial mathematics, finance and financial reporting, probability and mathematical statistics, models, contingencies, statistical methods, business economics, financial economics, and business awareness. The core application stage extends this knowledge by focusing more deeply on modeling and communication. The specialist technical stage then deals with health care, life insurance, general insurance, pensions and other benefits, finance and investment, general insurance (reserving and capital modeling and pricing), and enterprise risk management. These courses are followed by advanced study involving a dissertation-type research in one of the specialist fields.

Students may choose to complete an accredited actuarial science degree at an undergraduate or at a postgraduate level at a number of accredited universities and may thereby gain exemptions from certain professional examinations from the Institute. Most universities in the UK that offer actuarial science courses also require students to complete studies in various related fields including statistics, mathematics, applied mathematics, economics, and accounting.

Similar to the U.S. practice, Associates are members who have completed the core stage exams and have appropriate work experience.

Fellows are members of the Institute who have completed all examinations and meet the work experience requirement. Fellows and Associates have the right to call themselves actuaries.

The following website provides extensive information on the actuarial profession in the UK and affiliated countries: www.actuaries.org.uk/becoming-actuary/university-courses.

A significant number of local societies form the lifeline of the actuarial profession in the United Kingdom: Staple Inn Actuarial Society, Birmingham Actuarial Society, Bournemouth Actuarial Society, Bristol Actuarial Society, Channel Islands Actuarial Society, Faculty of Actuaries Students' Society, Glasgow Actuarial Students' Society, Invicta Actuarial Society, London Market Actuaries Group, London Market Students' Group, Manchester Actuarial Society, Manx Actuarial Society, Norwich Actuarial Society, Society of Actuaries in Ireland, White Horse Actuarial Society, and Yorkshire Actuarial Society.

The United Kingdom has reciprocity agreements with the countries belonging to the *Groupe Consultatif Actuariel Européen*. Some other countries, such as Hong Kong, use the UK process to certify actuaries.

Other Countries

The list of countries covered in this section is obviously incomplete. Most countries around the world have insurance companies and either privately run or public pension schemes. Actuarial considerations are therefore relevant to all countries. Rather than being encyclopedic, the choice of countries profiled in this section is intended to give you an idea of the variety of different national traditions and models for being an actuary. It also sheds some light on international mobility.

In his article on the state of actuarial science in China (see [2]), Alexander explains that China and other Pacific Rim countries are, at this point, *actuarially emerging countries*. We therefore refer to other sources for information about the state of the actuarial profession in these countries. A similar remark applies to Russia and many countries in Eastern Europe, the Middle East, Africa, and Central and South America. License to practice in emerging countries is often based on a mix of university education, working experience, and professional recognition by designated government agencies. Here are three typical examples from Eastern Europe:

In Croatia, an accredited actuary must be a full member of the Croatian Actuarial Association. To become accredited, candidates must have at least two years' actuarial work experience, and have successfully completed a program of actuarial study that follows the guidelines of the International Association of Actuaries and is recognized by the Assembly of the Croatian Actuarial Association.

An example of a recognized program of study is the Master's program in actuarial mathematics offered by the Department of Mathematics of the University of Zagreb, together with examinations set by the Croatian Actuarial Association.

Accredited actuaries must be Croatian citizens, have a degree in economics, mathematics, physics, or engineering, and have passed the examinations set by the Ministry of Finance or equivalent. An example of a recognized program of study is the actuarial program jointly organized by the Croatian Actuarial Association, the Department of Mathematics of Zagreb University, and the Actuarial Department of the Government of the United Kingdom.

To become accredited pension actuaries, candidates must first become accredited actuaries as described, have at least three to five years' experience in actuarial work (depending on their specialization at the undergraduate level), have a postgraduate degree in actuarial mathematics which includes pension insurance, or have an equivalent actuarial education from abroad, recognized by the Croatian Actuarial Association. In addition, they must have passed a special examination set by the Agency for Supervision of Pension Funds and Insurance.

In the Czech Republic, a permanent commission of the Czech Society of Actuaries, approved by the government, issues licenses to practice. Candidates must meet academic requirements based on the Czech tradition in actuarial education. The certification process also relies on the criteria for actuarial practice established by the *Groupe Consultatif*. In accordance with the Insurance Act, actuaries in the Czech Republic can become Appointed Actuaries upon approval by the Ministry of Finance. Foreign-trained actuaries must be fully qualified members of an actuarial society that is a full member of International Association of Actuaries.

In the Slovak Republic, anyone with an actuarially oriented university degree and one year of relevant work experience can become a member of the Slovak Society of Actuaries. To become an Appointed Actuary, Slovak insurance law requires that candidates have an appropriate university-level education, have passed a special examination, and have three years of relevant practical experience. The Slovak Financial Market Authority organizes the examination. Candidates are not required to be members of the Slovak Society of Actuaries.

Chapter 2

ACTUARIAL EDUCATION

All actuarial societies, be it the Society of Actuaries, the Canadian Actuarial Society, the Casualty Actuarial Society, the Institute and Faculty of Actuaries, the *Groupe Consultatif Actuariel Européen*, and others, expect their members to have passed a number of examinations in core courses. In this section, you will get a glimpse of what knowledge is involved.

2.1 | The International Syllabus

The International Actuarial Association surveyed its members about their educational practices. It identified a number of academic topics as describing the repertoire of scientific knowledge and competency areas of actuaries. The list of topics is, in a sense, more important than the answers to the survey. It summarizes the understanding of leading actuaries of the core tools of their profession. Here is the list, which divides actuarial education into 10 broad areas:

Financial Mathematics

Aim: To provide a grounding in the techniques of financial mathematics and their applications.

Topics: Introduction to asset types and securities markets; interest, yield, and other financial calculations; investment risk, introduction to stochastic interest and discount; market models, e.g., term structure of interest rates and cash flow models.

Probability and Mathematical Statistics

Aim: To provide a grounding in probability and mathematical statistics.

Topics: Concepts of probability; random variables and their characteristics; methods and properties of estimation; correlation and regression analysis; hypothesis testing and confidence intervals; data analysis.

Economics

Aim: To provide a grounding in the fundamental concepts of both microeconomics and macroeconomics.

Topics: Microeconomics; macroeconomics.

Accounting

Aim: To provide the ability to interpret the accounts and financial statements of companies.

Topics: Basic principles of accounting—including the role of accounting standards; different types of business entities; basic structure of company accounts; interpretation and limitation of company accounts.

Modeling

Aim: To provide an understanding of the principles of modeling and its applications.

Topics: Model structures; selection process; calibration; validation; scenario setting; sensitivity testing; limitations.

Statistical Methods

Aim: To provide the skills and expertise in the use of models appropriate for the understanding of risk in a range of actuarial work.

Topics: Statistical models, such as regression and time series; survival and multi-state models; risk models (individual and collective); parametric and nonparametric analysis of data; graduation principles and techniques; estimation of frequency, severity, and survival distributions; credibility theory; ruin theory.

Actuarial Mathematics

Aim: To provide the skills and expertise in the mathematics that are of particular relevance to actuaries working in life insurance, pensions, health care, and general insurance.

Topics: Actuarial mathematics as applied to life insurance, pensions, health care, and general insurance; types of products and plans—individual, group, and social insurance arrangements; pricing or financing methods of products and plans; reserving; reinsurance.

Investment and Asset Management

Aim: To develop the ability to apply actuarial principles to the valuation, appraisal, selection, and management of investments.

Topics: The objectives of institutional and individual investors; types of investment (bonds, shares, property, and derivatives); regulation and taxation of investments; valuation of investments; portfolio selection—incorporating assessment of relative value; performance measurement; portfolio management.

Principles of Actuarial Management

Aim: To develop the ability to apply the principles of actuarial planning and control needed for the operation of risk-related programs on sound financial lines.

Topics: The general operating environment; assessment of risks; product design and development; pricing and assumptions; reserving and valuation of liabilities; asset and liability relationships; monitoring the experience; solvency of the provider; calculation and distribution of profit (surplus).

Professionalism

Aim: To develop awareness of professionalism issues and the importance of professionalism in the work of an actuary.

Topics: Characteristics and standards of a profession; code of conduct and practice standards; the regulatory roles of actuaries; the professional role of the actuary.

This list can be found in the *Education Syllabus* section of the website of the International Actuarial Association. It is a wonderful conceptual organizer for the overwhelming mass of mathematical, economic, financial, and other ideas that make up the syllabus upon which the SAO and CAS examinations are based. As you read on, you might try to fit the listed topics and sample examination questions into this scheme. It will help you with the conceptual order and organization of the material that follows.

2.2 | The European Syllabus

The *Groupe Consultatif Actuariel Européen* was established in 1978 to bring together the actuarial associations in the European Union to represent the actuarial profession in discussions with European Union institutions on existing and proposed EU legislation that has an impact on the profession.

The *Groupe Consultatif* currently provides a forum for discussion amongst all actuarial associations throughout Europe; 35 actuarial associations from 33 European countries are represented. The *Groupe Consultatif* will organize its first European Congress of Actuaries in Brussels in 2012. The theme of the congress will be *The Future European Actuary.* Among its achievements is the formulation of a common syllabus for European actuaries. Not surprisingly, it also adheres to the four-corner structure. The syllabus was prepared in four stages:

- The preliminary stage that identified subjects "that are not unique to actuarial science but are essential background for study in this area." The subjects need not be covered individually but could be integrated with other subjects.

- The actuarial foundation stage that identified "subjects that form the fundamental tools for actuarial science and finance."

- The generalized applications stage that identified "subjects in which the principles and practice of actuarial techniques are developed in a variety of applications areas."

- The country specific and specialist stage that identified "the detailed regulatory, legislative, cultural and administrative framework of the country in which they intend to work."

The syllabus also deals with the post-qualification training necessary to ensure that actuaries are up-to-date with changes in the framework for their practice area and includes a continuing professional development (CPD) component.

The subjects covered in the syllabus include mathematics, probability and statistics, stochastic processes, computing, economics, accounting and financial reporting, structures and legislative instruments of the European Union, communication skills, and language skills. The website of the *Groupe* describes in detail all of the academic and professional components of the syllabus. Even a cursory reading of the SOA, CAS, and *Groupe* syllabi shows that there is remarkable agreement among actuaries in the countries concerned regarding what academic and profession skills actuaries should have and should continue to develop for professional competency and lifelong success.

The details of the *Groupe*'s syllabus are available for downloading on the Groupe's website: www.gcactuaries.org/documents/core_2012.pdf.

2.3 The North American Syllabus

There is an old saying: *The more things change, the more they stay the same.* Over the last ten years, the organization of the actuarial examinations and distribution of topics over the underlying courses has changed. Some topics were reorganized, some examinations are now online, and professional and experiential components have been strengthened. Nevertheless, the mathematical underpinnings of the profession are essentially the same, now and then. Since the survey on which this guide is based was conducted almost ten years ago, the respondents' comments on specific actuarial courses and examinations refer to the system that was in place at that time. However, the core of this book consists of the wisdom and experience expressed in the results of a survey, which readers found both stimulating and informative. It was therefore essential in preparing this second edition of this book to retain the flavor of survey answers. This made sense since, as explained in the Preface, the content of the examinations has not changed substantially. Only the numbers, letters, and names of the courses have changed and some of the material has been organized in modular form. In most cases the context makes it clear which courses and examinations are at play.

In the "Actuaries around the World" section in Chapter 1, we saw that there are major differences in actuarial education around the world. However, the SOA and CAS qualifications are respected and honored around the world. Students can write SOA and CAS examinations in multiple centers in the United States and most provinces in Canada, as well as in international examination centers. If you peruse the SOA, CAS, and CIA websites, you will see a long list of biannual American and Canadian examination centers and the dates of the examinations. But the list of international centers is equally impressive.

If you are lucky enough to find international employment, you might want to enhance your actuarial credentials by taking some of your examinations abroad. Here is the list of permanent international test centers outside the United States and Canada, where you are able to write SOA and CAS examinations: Accra (Ghana), Athens (Greece), Bangkok (Thailand), Beijing (China), Bogota (Colombia), Bridgetown (Barbados), Buenos Aires (Argentina), Cairo (Egypt), Cape Town (South Africa), Changsha (China), Colombo (Sri Lanka), Delhi (India), Guangzhou (China), Hamilton (Bermuda), Harare (Zimbabwe), Hefei (China), Ho Chi Minh City (Vietnam), Hong Kong, Hyderabad (India), Jakarta (Indonesia), Johannesburg (South Africa), Karachi (Pakistan), Kingston (Jamaica), Kuala Lumpur (Malaysia), Lagos (Nigeria), Lahore (Pakistan), Madrid (Spain), Manila (Philippines), Mumbai (India), Nairobi (Kenya), Nassau (Bahamas), Oxford (England), Panama City (Panama), Paris (France), Port-of-Spain (Trinidad), Ramat Gan (Israel), Santiago (Chile), São Paulo (Brazil), Seoul (South Korea), Shanghai (China), Shenzhen (China), Singapore, Sydney (Australia), Taichung (Taiwan), Taipei (Taiwan), Tianjin (China), Tokyo (Japan), Warsaw (Poland), Xi'an (China), and Zurich (Switzerland).

This list certainly shows you the high regard in which the North American system of actuarial education is held around the world. In places where no permanent examination centers exist, arrangements can often by made by candidates by finding their own supervisors of examinations acceptable to the Society of Actuaries. Supervisors must be either members of the Society of Actuaries, members of the Casualty Actuarial Society, members of the Institute and Faculty of Actuaries (United Kingdom), or be a tenured academic or other qualified testing professional. If no such supervisor is available, approval may even be given for writing the examinations at an Embassy of the United States.

2.4 | The SOA Examinations

The Society of Actuaries currently offers three pathways to professional accreditation: Associateship (ASA), Fellowship (FSA), and Chartered Enterprise Risk Analyst (CERA). The website of the Society provides an interactive survey of the requirements:

Associateship Examinations

- FAP
- Exam P, Exam MFE, Exam C, and Exam APC
- Exam FM, Exam MLC
- VEE

According to the Society, *an Associate of the Society of Actuaries has demonstrated knowledge of the fundamental concepts and techniques for modeling and managing risk. The Associate has also learned the basic methods of applying those concepts and techniques to common problems involving uncertain future events, especially those with financial implications.* Candidates for Associateship are required to demonstrate that they possess this knowledge by passing a number of required courses. The Society describes these course components as follows:

- FAP (Fundamentals of Actuarial Practice) consists of an e-learning course and written assessments and can be completed anytime over a two-year period. The course comprises eight modules:
    - Module 1: The Role of the Professional Actuary
    - Module 2: Core External Forces
    - Module 3: Risk in Actuarial Problems
    - Module 4: Actuarial Solutions
    - Module 5: Actuarial Models
    - Module 6: Model Selection and Solution Design
    - Module 7: Selection of Initial Assumptions
    - Module 8: Monitoring Results and Completing the Control Cycle
- Exam P (Probability): According to the Society, *the syllabus for Exam P develops the candidate's knowledge of the fundamental probability tools*

for quantitatively assessing risk. The application of these tools to problems encountered in actuarial science is emphasized. A thorough command of the supporting calculus is assumed. Additionally, a very basic knowledge of insurance and risk management is assumed.

- Exam FM (Financial Mathematics): The syllabus for Exam FM is designed to develop *the candidate's understanding of the fundamental concepts of financial mathematics, and how those concepts are applied in calculating present and accumulated values for various streams of cash flows as a basis for future use in reserving, valuation, pricing, asset/liability management, investment income, capital budgeting and valuing contingent cash flows. The candidate will also be given an introduction to financial instruments, including derivatives, and the concept of no-arbitrage as it relates to financial mathematics. A basic knowledge of calculus and an introductory knowledge of probability is assumed.*

- Exam MFE (Models for Financial Economics): The syllabus for Exam MFE *develops the candidate's knowledge of the theoretical basis of certain actuarial models and the application of those models to insurance and other financial risks. A thorough knowledge of calculus, probability and interest theory is assumed.*

- Exam MLC (Models for Life Contingencies): The syllabus for Exam MLC *develops the candidate's knowledge of the theoretical basis of certain actuarial models and the application of those models to insurance and other financial risks. A thorough knowledge of calculus, probability and interest theory is assumed.*

- Exam C (Construction and Evaluation of Actuarial Models): The syllabus for Exam C *provides an introduction to modeling and covers important actuarial methods that are useful in modeling. A thorough knowledge of calculus, probability and interest theory is assumed.*

- APC (Associateship Professionalism Course): The APC is a one-day seminar covering professionalism, ethics and legal liability and uses case studies.

- VEE (Validation by Educational Experience): A candidate passes the VEE component of accreditation *when the candidates show that they have completed the coursework in specific subjects. After completing any two exams, candidates may apply to have the VEE credit earned applied to their records. The VEE requirements are completed by receiving a specified grade on VEE-approved college classes, online courses or standard examinations. The VEE subjects consist of Economics, Corporate Finance and Applied Statistics.*

The SOA website describes in detail the examination and accreditation process. You can also become an Associate by becoming a Certified Enterprise Risk Analyst (CERA), an alternative career pathway described below.

Fellowship SOA Examinations

According to the Society, *Fellows of the Society of Actuaries must have demon-strated knowledge of the business environments within which financial decisions concerning retirement benefits, life insurance, annuities, health insurance, invest-ments, finance, and enterprise risk management are made, including the appli-cation of advanced concepts and techniques for modeling and managing risk.* In addition, Fellows *must have in-depth knowledge of the application of appropri-ate concepts and techniques to a specific area of actuarial practice.* Candidates for Fellowship must demonstrate that they possess this knowledge by passing a number of required courses in specific tracks. The Society describes these tracks as follows:

When you are ready to take the Fellowship requirements, you must select a speciality track and complete all requirements in that track. (Mixing requirements from different tracks is not permitted.) The Society offers five speciality tracks: Finance/ERM, Individual Life and Annuities, Group and Health, Investment, and Retirement Benefits. The description of these five tracks give you a very good idea of the mainstream activities of the members of the Society.

Finance/ERM Track

The SOA syllabus states that *this track consists of five components includ-ing exams, e-learning modules and a 3-day seminar. After completing the Finance/ERM track, candidates will have a thorough understanding of the appli-cations of economic theory to financial markets and modeling, advanced financial issues, and enterprise risk management. The track is designed to prepare candi-dates for employment in the fields of finance and enterprise risk management.*

- The Financial and Health Economics Module *focuses on how individuals acquire, save and invest money. The module covers this in general as well as the specific cases of the financial services industry and health care delivery systems. The Financial and Health Economics FSA Module is an e-learning module where candidates acquire and use knowledge that is distributed and facilitated electronically. Financial Economics is the discipline underlying all financial services.*

- The Financial Reporting Module *exposes candidates to the concepts and ter-minology necessary to understand financial statements and regulatory capital requirements. The Financial Reporting FSA Module is an e-learning module where candidates acquire and use knowledge that is distributed and facili-tated electronically.*

- The Operational Risk Module *prepares candidates to improve their practice of operational risk and places it in the context of enterprise risk management. The Operational Risk Module is an e-learning module where candidates acquire and use knowledge that is distributed and facilitated electronically.*

- The syllabus for the Financial Economic Theory and Engineering Exam *covers applications of economic theory to financial markets and modeling. The candidates are exposed to modern corporate finance theory and applications, advanced derivatives and financial markets pricing and modeling, and information theory and efficient markets theory.*

- The syllabus for the Advanced Finance/ERM Exam *covers advanced financial issues and enterprise risk management. The candidates are exposed to methods for risk identification, quantification and valuation, and techniques for risk management.*

- The DMAC e-Learning Module is one part of the Capstone Experience—the final component prior to Fellowship. According to the Society, the Capstone Experience *includes the DMAC module and the Fellowship Admissions Course (FAC). DMAC is an e-learning module where the candidate acquires and uses knowledge that is distributed and facilitated electronically. The electronic e-learning facility will also provide information regarding any required text readings and other non-computer–based activities. There is also a project component. The focus of the Decision Making and Communication Module is on written and oral communication skills and decision-making skills as applied to solving business problems.*

- The final requirement to attain the Fellowship designation is the Fellowship Admissions Course (FAC). As stated by the Society, *the FAC can be taken any time after all other requirements have been completed. Successful completion of the course requires attendance and participation in all education sessions as well successful completion of the oral presentation requirement.*

Individual Life and Annuities Track

This track consists of five components including exams, e-learning modules, and a 3-day seminar. After completing the Individual Life and Annuities track, candidates will have a thorough understanding of individual life and annuity products. The aspects covered are: financial reporting, valuation, product design, and pricing. The track is designed to prepare candidates for employment in the fields of financial reporting and product development.

- The Regulation and Taxation FSA Module is *an e-learning module where you acquire and use knowledge that is distributed and facilitated electronically. The electronic e-learning facility will also provide information regarding required text readings and other non-computer–based activities. The goal of the Regulation and Taxation Module is to provide you with an understanding of, and appreciation for, the regulatory environment that affects the life and annuity insurance industry overall and actuaries in particular.*

- The Operational Risk Module or Financial Reporting Module described above.

- The Individual Life and Annuities DP Exam *is a written-answer examination offered in the spring and fall of each year. Both a U.S. and Canadian version of the exam are administered, each of which has its own reading list. The syllabus for this exam covers the individual life insurance and annuity product design and development process. The course of reading exposes the candidate to the relationship between the product design and the selection of appropriate pricing assumptions as well as the profit measures used in pricing.*

- The Individual Life and Annuities CSP *is a written-answer examination offered in the spring and fall of each year. Both a U.S. and Canadian version of the exam are administered, each of which has its own reading list. The syllabus for this exam exposes the candidate to the preparation and analysis of life insurance company financial statements including the underlying valuation principles supporting these statements. Further, the course of reading covers methods of reinsurance, risk mitigation, determining risk based and economic capital, asset liability matching and calculating embedded value for individual life insurance and annuity products.*

- The DMAC e-Learning Module described above.

- The Fellowship Admissions Course (FAC) described above.

Group and Health Track

This track consists of exams, e-learning modules, and a 3-day seminar. According the the SOA syllabus, *after completing the Group and Health track, candidates will have a thorough understanding of group and health products. The aspects covered are: financial reporting, valuation, product design, and pricing. The track is designed to prepare candidates for employment in the fields of financial reporting, product development, and pricing.*

- The Financial and Health Economics Module described above.

- The Health Foundations Module *is an e-learning module where you acquire and use knowledge that is distributed and facilitated electronically. The module discusses the health care system at a micro level. Through an understanding of terminology and coding, health care data and research can be understood, effectively assessed for quality, and effectively used in actuarial work.*

- The Pricing, Reserving and Forecasting Module *is an e-learning module where you acquire and use knowledge that is distributed and facilitated electronically. In this module candidates will be exposed to practical techniques involved in managing the financial control cycle of a health care company—from trend determination to pricing and reserving to analysis of historical results to forecasting future experience.*

- The Group and Health DP *is a written-answer examination offered in the spring and fall of each year. The syllabus for this exam develops the candidate's knowledge of the design and pricing of group and health products. Specific topics include: single employer group coverage, individual and multi–life coverages, coverages offered through group vehicles, employer strategies for design and funding, government health plans, regulation, taxation, utilization and claims management, predictive modeling techniques, medical manual rates, credibility theory, and underwriting.*

- According to the SOA syllabus, *the Group and Health CSP Exam is a written-answer examination offered in the spring and fall of each year. The syllabus for this exam develops the candidate's knowledge of the financial analysis of group and health products including the underlying valuation principles. Specific topics include: quality measures, markets, legal requirements, reserving techniques for claims and other liabilities, financial performance measures, reinsurance arrangements, company taxation, regulations, risk profile of product portfolios, capital needs assessments, actuarial opinions, and actuarial appraisals.*

- The DMAC e-Learning Module described above.

- The Fellowship Admissions Course (FAC) described above.

Investment Track

This track consists of exams, e-learning modules, and a 3-day seminar. After completing this track, you will have a thorough understanding of portfolio management and the applications of economic theory to financial markets and modeling. The track is designed to prepare candidates for employment in the fields of investments and portfolio management.

- According to the SOA, the Advanced Portfolio Management exam *is a written-answer examination offered in the spring and fall of each year. The syllabus for this exam covers the portfolio management process in detail. The candidate is exposed to the investment management process, types of asset classes, techniques for managing equity, fixed income, and alternative asset portfolios, credit risk modeling and management, performance measurement, and behavioral aspects of portfolio management.*

- The Financial Economic Theory and Engineering exam *is a written-answer examination offered in the spring and fall of each year. The syllabus for this exam covers applications of economic theory to financial markets and modeling. The candidate is exposed to modern corporate finance theory and applications, advanced derivatives and financial markets pricing and modeling, and information theory and efficient markets theory.*

- The Financial and Health Economics Module described above.

- The Investment Strategy Module *is an e-learning module where you acquire and use knowledge that is distributed and facilitated electronically. The electronic e-learning facility will also provide information regarding required text readings and other non-computer–based activities. The focus of the Investment Strategy Module is to provide you with an understanding of the investment theories to implement the investment process.* The goal of the module is to enable you to construct and recommend portfolios based on given objectives.

- The Operational Risk Module or Financial Reporting Module described above.

- The DMAC e-Learning Module described above.

- The Fellowship Admissions Course (FAC) described above.

Retirement Benefits Track

After completing the Retirement Benefits track, you will have a thorough understanding of retirement plans. The aspects covered are plan design, funding, accounting, and valuation. This track prepares you for employment in the fields of defined benefit plans, defined contribution plans, and retiree health plans.

- As stated by the SOA, the *Retirement Benefits CSP is a written-answer examination offered in the spring and fall of each year. Both a U.S. and Canadian version of the exam are administered, each of which has its own reading list. The syllabus for this exam develops the candidate's ability to analyze retirement plan design, valuation, funding and accounting, both in general and as it applies to a particular situation.*

- The Retirement Benefits DP *is a written-answer examination offered in the spring and fall of each year. Both a U.S. and Canadian version of the exam are administered, each of which has its own reading list. The syllabus for this exam covers the basics of design and valuation for the full gamut of retirement plans found in the U.S. and Canada. Specific topics include qualified defined benefit and defined contribution plans, retiree health plans, nonqualified plans, international plans, actuarial valuation methods and assumptions, regulation, plan investments, and professional standards of practice.*

- The Financial and Health Economics Module described above.

- The Social Insurance Module *is an e-learning module where you acquire and use knowledge that is distributed and facilitated electronically. The electronic e-learning facility will also provide information regarding required text readings and other non-computer–based activities. Employer-provided benefits should be designed to coordinate with social insurance programs. As you will learn in this module, the social insurance programs vary widely in their coverage, cost and benefits in different countries. It is critical that you be prepared to coordinate the employer-provided plans with the social insurance programs in your country. The Social Insurance Module goes into the detail needed to practice as an actuary in the United States or Canada.*

- The Operational Risk Module *is an e-learning module where you acquire and use knowledge that is distributed and facilitated electronically. The electronic e-learning facility will also provide information regarding required text readings and other non-computer–based activities. The purpose of the Operational Risk Module is to expose you to the concepts and terminology necessary to understand operational risk. With the increasingly complex and fast changing business environment, organizations are seeking risk management professionals to join their teams.*

- The Investment Strategy Module *is an e-learning module where you acquire and use knowledge that is distributed and facilitated electronically. The electronic e-learning facility will also provide information regarding required text readings and other non-computer–based activities. The focus of the Investment Strategy module is to provide you with an understanding of the investment theories to implement the investment process. Throughout the module, you will be exposed to case studies from real experiences that illustrate the range of considerations in managing investment portfolios supporting particular liabilities and goals. The primary goal of the module is to enable you to construct and recommend a portfolio based on given objectives.*

- The SOA administers multiple-choice examinations on behalf of the Joint Board for the Enrollment of Actuaries. The basic EA-1 and the pension EA-2 (Segment B) examinations are administered in the spring of each year. The pension EA-2 (Segment A) examination is administered in the fall. The course of reading for the basic EA-1 examination covers mathematics of compound interest, mathematics of life contingencies, and demographic analysis. The course of reading for the pension EA-2 examination (Segment A) covers actuarial assumptions, actuarial cost methods, calculation of minimum required contributions, and calculation of maximum deductible contributions. The course of reading for the pension EA-2 examination (Segment B) covers ERISA and related laws. Both Segment A and Segment B of pension EA-2 are required for the SOA Retirement Benefits (U.S. version) specialty track for Fellowship.

- The DMAC e-Learning Module described above.

- The Fellowship Admissions Course (FAC) described above.

Chartered Enterprise Risk Analyst Examinations

As is shown on the SOA website, *a new designation—the Chartered Enterprise Risk Analyst (CERA)—is now available to help students and business professionals prepare for and seize opportunities in the evolving discipline of enterprise risk management (ERM) within broader financial services, insurance, and pension markets.*

According to the SOA, *the syllabus was developed to meet market needs and provides a rigorous treatment of critical ERM topics, including actuarial approaches to risk. It is important to note that successful candidates receive the Chartered Enterprise Risk Analyst (CERA) designation and at the same time are accredited as Associates of the Society of Actuaries (ASA).* The required examinations are offered by SOA and consist of the following:

- Exam P (Probability)

- Exam FM (Financial Mathematics)

- Validation by Educational Experience (VEE) Economics

- Exam M (Actuarial Models) segment MFE

- Exam C (Construction of Actuarial Models)

- FSA-level Finance/ERM Exam (Advanced Finance/Enterprise Risk Management)

- FSA-level Finance/ERM Module (Financial Reporting and Operational Risk–FSA Module Introduction, Operational Risk Section and End-of-Module Exercise only)

- The Associateship Professionalism Course

Continuing Professional Development

One of the important professional innovations introduced by the Society of Actuaries (SOA) is the Continuing Professional Development (CPD) Requirement for its membership. According to the Society, *all members must demonstrate on a regular basis that they are "SOA Compliant" by having met the SOA CPD Requirement during the past two years, a requirement that began December 31, 2010.*

What this means in practice is discussed in detail on the SOA website. According to the Society, *members who are not compliant failed to earn sufficient continuing education credits to be able to attest compliance, elected not to comply with the requirement, or did not attest compliance.*

The Society states that *members do not lose their SOA designations (ASA, CERA, FSA) because they fail to comply with the SOA CPD Requirement (although they can lose their designation if they have been found to violate the terms of non-compliance).* On the other hand, *all non-compliant members are required to tell users of their actuarial expertise of their failure to comply with the SOA CPD Requirement.*

The compliance component obviously places considerable ongoing importance, therefore, on one of the four cornerstones of the actuarial profession mentioned earlier: core knowledge, specialized knowledge, *professional knowledge*, and experiential knowledge.

2.5   The CAS Examinations

As in the case of SOA, the CAS syllabus has four corners: core knowledge, specialist knowledge, professional knowledge, and experiential knowledge. Accreditation proceeds at two levels: candidates become Associates if they meet the core and some specialist requirements, as well as some professional and experiential requirements. Let us take a closer look at the examinations involved.

Associateship Examinations

Exam 1—Probability

The Probability Exam *is a three-hour multiple-choice examination. It is called Exam P by the Society of Actuaries and Exam 1 by the Casualty Actuarial Society. The Probability Exam is offered via computer-based testing (CBT).*

The syllabus for the Probability Exam develops the candidate's knowledge of the fundamental probability tools for quantitatively assessing risk. The application of these tools to problems encountered in actuarial science is emphasized. A thorough command of the supporting calculus is assumed. Additionally, a very basic knowledge of insurance and risk management is assumed.

Exam 2—Financial Mathematics

The Financial Mathematics Exam *is a three-hour multiple-choice examination. It is called Exam FM by the Society of Actuaries and Exam 2 by the Casualty Actuarial Society. The financial mathematics exam is offered via computer-based testing (CBT).*

The syllabus for the Financial Mathematics Exam develops the candidate's understanding of the fundamental concepts of financial mathematics and how those concepts are applied in calculating present and accumulated values for various streams of cash flows as a basis for future use in: reserving, valuation, pricing, asset/liability management, investment income, capital budgeting, and valuing contingent cash flows. The candidate will also be given an introduction to financial instruments, including derivatives, and the concept of no-arbitrage as it relates to financial mathematics. A basic knowledge of calculus and an introductory knowledge of probability is assumed.

Exam 3F—Models for Financial Economics

The Models for Financial Economics Exam *is a three-hour multiple-choice examination. It is called Exam MFE by the Society of Actuaries and Exam 3F by the Casualty Actuarial Society. This exam is offered via computer-based testing (CBT).*

The syllabus for the Models for Financial Economics Exam develops the candidate's knowledge of the theoretical basis of certain actuarial models and the application of those models to insurance and other financial risks. A thorough knowledge of calculus, probability and interest theory is assumed. Knowledge of risk management at the level of Exam 1/P is also assumed. In addition, for Exam MFE/3F, candidates are assumed to be familiar with the earlier chapters of the McDonald text, which are in the syllabus of Exam 2/FM.

Exam 3L—Models for Life Contingencies and Statistics

Exam 3L *is a two-and-a-half-hour, multiple-choice exam on life contingencies and statistics that is administered by the CAS. This material develops the candidate's knowledge of the theoretical basis of certain actuarial models and the application of those models to insurance and other risks. A thorough knowledge of calculus, probability and interest theory is assumed. Knowledge of risk management at the level of Exam 1/P is also assumed.*

This examination develops the candidate's knowledge of the theoretical basis of contingent payment models and the application of those models to insurance risks.

The candidate will be required to develop an understanding of contingent payment models. The candidate will be expected to understand what important results can be obtained from these models for the purpose of making business decisions, and what approaches can be used to determine these results.

Exam 4—Construction and Evaluation of Actuarial Models

The Construction and Evaluation of Actuarial Models Exam *is a three-and-a-half hour multiple-choice examination. It is called Exam C by the Society of Actuaries and Exam 4 by the Casualty Actuarial Society. It is offered via computer-based testing (CBT).*

The syllabus provides an introduction to modeling and covers important actuarial methods that are useful in modeling. A thorough knowledge of calculus and probability is assumed.

The candidate will be introduced to useful frequency and severity models beyond those covered in the Financial Economics and Life Contingencies Exams. The candidate will be required to understand the steps involved in the modeling process and how to carry out these steps in solving business problems. The candidate should be able to analyze data from an application in a business context; determine a suitable model including parameter values; and provide measures of confidence for decisions based upon the model. The candidate will be introduced to a variety of tools for the calibration and evaluation of the models.

Exam 5—Basic Techniques for Ratemaking and Estimating Claim Liabilities

Before commencing study for this four-hour examination, candidates should read the "Introduction" to "Materials for Study" for important information about learning objectives, knowledge statements, readings, and the range of weights. Items marked with a bold SK constitute the 2011 CAS Exam 5 Study Kit that may be purchased from the CAS Online Store. Items marked with a bold OP (Online Publication) or W (the 2011 Exam 5 Web Notes) are available at no charge and may be downloaded from links in the Complete Text References section below—or a printed version may be purchased from the CAS Online Store.

Exam 6—Regulation and Financial Reporting (Nation-Specific)

- **(Taipei)**

  The Taipei version of this examination is based on the CAS examinations for property casualty actuaries. The CAS Board of Directors actually approved the Taipei-specific examinations as *fulfilling the nation specific requirements for CAS member effective January 1, 2010.*

- **(Canada)**

  The Canadian version of this four-hour examination assumes that candidates have completed the Online Course 2. This course contains the fundamental material for both the Regulations of Insurance and Canadian Insurance Law, as well as the Canadian financial reporting requirements.

  The course description states that *the examination will test the candidate's knowledge of topics that are presented in the learning objectives. Thus, the candidate should expect that each exam will cover a large proportion of the learning objectives and associated knowledge statements and syllabus readings, and that all of these will be tested at least once over the course of a few years but each one may not be covered on a particular exam.*

- **(United States)**

  The syllabus of the U.S. version of this course states that *Section A of this examination covers insurance regulation with regards to property-casualty coverages, ratemaking, pricing, and solvency, and U.S. tort law as it affects the property-casualty business. Section B covers markets, coverages, and private and governmental programs for the property-casualty business in the United States. Section C covers the aspects of statutory, Generally Accepted Accounting Principles (GAAP), and International Financial Reporting Standards (IFRS) insurance accounting and taxation as these affect reserving and statutory reporting in the United States. Section D covers the professional responsibilities of the appointed actuary according to the Property and Casualty Annual Statement Instructions issued by the National Association of Insurance Commissioners (NAIC). Section E presents the general concepts of reinsurance accounting to the candidate.*

VEE Requirement

The VEE requirement of CAS is identical to that of the Society of Actuaries and the Canadian Institute of Actuaries. It again consists of three topics: Economics, Corporate Finance, and Applied Statistical Methods. As the CAS website indicates that *the VEE topics are not prerequisites for the preliminary exams (Exams P/1, FM/2, MLC, MFE/3F and C/4) and may be fulfilled independently of the exam process. However, you must pass two SOA or CAS actuarial exams before applying to have your VEE credit added to your record.*

Professionalism Course

The CAS also has a professionalism requirement. The CAS website states that *to be eligible for the CAS course on professionalism, a candidate must have either credit for any four actuarial exams and credit for any four of the following five requirements: (Online Course 1, Online Course 2, VEE-Applied Statistical Methods, VEE-Corporate Finance, or VEE-Economics), or credit for any five actuarial exams in the current education structure—regardless of Internet courses or VEE status.*

Fellowship Examinations

Exam 7—Estimation of Policy Liabilities, Insurance Company Valuation, and Enterprise Risk Management

This course tests "a candidate's knowledge of topics that are presented in the learning objectives. Thus, the candidate should expect that each exam will cover a large proportion of the learning objectives and associated knowledge statements and syllabus readings, and that all of these will be tested at least once over the course of a few years—but each one may not be covered on a particular exam."

Exam 8—Advanced Ratemaking

This course is designed to test a "candidate's knowledge of topics that are presented in the learning objectives. Thus, the candidate should expect that each exam will cover a large proportion of the learning objectives and associated knowledge statements and syllabus readings, and that all of these will be tested at least once over the course of a few years—but each one may not be covered on a particular exam."

"Candidates for Exam 8 are expected to have already acquired considerable technical knowledge and practical experience in insurance ratemaking. Therefore, this examination will assume a working knowledge of basic ratemaking and will deal with advanced topics. To some degree, the examination will deal with the types of practical problems that a fully qualified actuary, working in ratemaking, should be able to solve. The ability to apply ratemaking knowledge and experience may be tested through questions dealing with problems for which there are no generally recognized solutions."

Exam 9—Financial Risk and Rate of Return

This examination tests a *candidate's knowledge of topics that are presented in the learning objectives. Thus, the candidate should expect that each exam will cover a large proportion of the learning objectives and associated knowledge statements and syllabus readings, and that all of these will be tested at least once over the course of a few years—but each one may not be covered on a particular exam.*

Exam 9 focuses on a broad array of finance, investment, and financial risk management topics. This examination assumes a working knowledge of basic ratemaking, finance, probability and statistical modeling, liability and reserve risk, and insurance underwriting. The ability to apply this knowledge and experience may be tested through questions dealing with problems for which there are no generally recognized solutions.

2.6 Printed Study Material

The Actuarial Bookstore sells comprehensive sets of actuarial examination tools, for both the SOA and CAS examinations. For details, see the bookstore website at www.actuarialbookstore.com/exam/listing.asp. By browsing through the list of available tools, you can get a sense, from another perspective, of the range of career options available to budding actuaries:

SOA Examinations

- SOA Exam P/CAS Exam 1—Probability
- SOA Exam FM/CAS Exam 2—Financial Mathematics
- SOA Exam MLC—Life Contingencies
- SOA Exam MFE/CAS Exam 3F—Financial Economics
- SOA Exam C/CAS Exam 4—Construction and Evaluation of Actuarial Models
- SOA FAP—Fundamentals of Actuarial Practice
- SOA CSP—Individual Life and Annuity
- SOA CSP—Health, Group Life and Managed Care
- SOA Advanced Portfolio Management
- SOA Financial Economic Theory
- SOA Advanced Finance/ERM (Enterprise Risk Management)
- SOA DP—Health, Group Life and Managed Care
- SOA DP—Individual Life and Annuity
- SOA DP—Retirement (U.S. and Canada)

FSA Modules

- FSA Module—Financial and Health Economics (1): for Finance/ERM, Investment, Individual Life and Annuities, and Retirement Benefits track candidates
- FSA Module—Financial and Health Economics (2): for Group and Health track candidates
- FSA Module—Financial Reporting
- FSA Module—Health Foundations
- FSA Module—Investment Strategy
- FSA Module—Operational Risk: for Finance/ERM, Investment, and Individual Life and Annuities track candidates selecting Operational Risk
- FSA Module—Social Insurance

- FSA Module—Pricing, Reserving and Forecasting
- DMAC—Decision Making and Communication Module
- DMAC—Decision Making and Communication Module (2011 Redesign)

CAS Examinations

- SOA Exam P/CAS Exam 1—Probability
- SOA Exam FM/CAS Exam 2—Financial Mathematics
- CAS 3L—Life Contingencies
- SOA Exam MFE/CAS Exam 3F—Financial Economics
- SOA Exam C/CAS Exam 4—Construction and Evaluation of Actuarial Models
- CAS Exam 5—Basic Techniques for Ratemaking and Estimating Claim Liabilities
- CAS Exam 6—Regulation and Financial Reporting
- CAS Exam 7—Estimation of Policy Liabilities, Insurance Company Valuation, and Enterprise Risk Management
- CAS Exam 8—Advanced Ratemaking
- CAS Exam 9—Financial Risk and Rate of Return

SOFE Exams

- AFE—Accredited Financial Examiner
- CFE—Certified Financial Examiner

EA Exams

- EA-1—Basic Examination
- EA2-A—Pension Examination (A)
- EA2-B—Pension Examination (B)

VEE Exams

- VEE Economics
- VEE Corporate Finance
- VEE Applied Statistical Models

Seminars and Online Courses

- SOA Seminars
- CAS Seminars
- VEE On Line Courses

CERA Designation

- VEE Economics
- SOA Exam P/CAS Exam 1—Probability
- SOA Exam FM/CAS Exam 2—Financial Mathematics
- SOA Exam MFE/CAS Exam 3F—Financial Economics
- SOA Exam C/CAS Exam 4—Construction and Evaluation of Actuarial Models
- SOA Advanced Finance/ERM
- FSA Module—Operational Risk: for Finance/ERM, Investment, and Individual Life and Annuities track candidates selecting Operational Risk

General Comments

If you take a closer look at the first four SOA and CAS courses, you will discover that most of them deal with some aspect of mathematics and statistics in a business context. In addition to being able to carry out mathematical and statistical calculations, you must be able to understand the definitions and interrelationships of concepts from economics and finance. You will notice that many questions involve both definite, indefinite, and multiple integrals, as well as ordinary and partial derivatives. Hence, a good command of calculus is essential. Exponential and logarithmic functions are core functions, and so are geometric progressions. You will also discover that indispensable concepts from statistics are probabilities, distributions, random variables, expected values, mean, and variance. You will also notice that among the probability distributions, the Poisson distribution comes up most often. While the specific questions will obviously vary from year to year, it is clear from the nature of actuarial science that the mentioned ideas and techniques from mathematics, business, and statistics will always be part of the skills an actuary is required to possess.

| 2.7 | **Theory and Practice** |

Q What is the connection between the actuarial examinations and the knowledge and skills required in actuarial practice?

Answer Not as important as I expected, especially the first four exams, which are very different from actuarial practice and so are the required skills. The examinations in Courses 5 [2002] and 6 [2002] are closer to the real world, and I have heard that the examinations in Courses 7 [2002] and 8 [2002] are much more like real consulting situations, although you can't really have a consulting situation in an exam. My overall feeling is that the more you advance in your exams, the more related they are to real life. I just don't feel that the mathematics background for the examinations in Courses 1 [2002] through 4 [2002] is that important in real life. I am not saying it is not important, just that it is not a major part of success in the workplace in the first few years of employment.

Answer Limited. Material is either too theoretical, or off whatever is required from us in the day-to-day life.

Answer I would say that the actuarial examinations go much deeper. We have to know every little formula, even if it applies only in a unusual situation. In our job, we can use one formula and adapt it to given situations. Also, the Associateship exams touch every field (pension, insurance, finance, etc.). So an actuary needs to be versatile.

Answer Actuaries need to follow specific methods that are regulated in order to make sure everyone follows the same standards. The actuarial examinations are a way to introduce the various methods of calculating reserves, and they give you a background in subjects used in real life situations.

Answer Actuarial exams helped me to develop my capacity to focus on problems and to solve them. Since I can solve a lot of problems, I do not need to constantly refer to books. Also, actuarial examinations help us learn how to work really hard and well.

Answer The knowledge acquired in the study for actuarial examinations serves as a good learning base for the actuarial practice. As in all professions, most of the learning is done on the job, but in the case of actuaries, since such an extensive knowledge must be acquired, the examinations form the practicing actuary.

Answer Courses 5 [2002] to 8 [2002] are more directly applicable to work, although Courses 2 [2002] and 3 [2002] are often directly used too. A strong basis in the fundamentals of Courses 1 through 4 is needed to do Courses 5 [2002] through 8 [2002]. Some topics only apply to certain jobs and the degree of applicability is job-specific too. I always treated Courses 1 [2002] through 4 [2002] as first- or second-year university courses. This puts students from various educational backgrounds on the same common basis from which they can build up their knowledge. Courses 5 [2002] through 8 [2002] are more like third- or fourth-year university courses—higher level of knowledge, more practical, some specialization occurring.

Answer The mathematical exams (Courses 1 [2002] through 4 [2002]) are directly related to the actuarial needs. Other exam material is necessary, but the way it is tested (i.e., learning everything by heart) is not related to actuarial needs.

Answer Not much when it comes to the learning things by heart part, because you have all the documentation handy, especially with the net. The first part (mathematics), though, is really good.

Answer I see the applications of life contingencies daily in my pension work.

Answer The actuarial exams expose the student to some of the available literature in the field. The actuary may refer back to some of these sources later.

Answer Actuarial examinations provide a very technical introduction to the skills required in actuarial practice. Most of the formulas memorized in the early mathematical exams will hardly be used in practice. Knowing how to apply the concepts presented in the examinations along with the skills outlined in the previous question are really what are required in actuarial practice.

Answer Actuarial examinations will teach the actuary all the required actuarial skills. They are necessary. They will also teach (at least for CAS exams) about the insurance business. But they will not teach business skills. Basically, on the P/C side, exams teach how to price products, how to calculate IBNR (incurred but not reported) loss reserves, how to read a financial statement, how to value investments, how to perform modeling, and also teach the basics of P/C insurance (products, concepts, etc.). I would say all exams are relevant to actuaries.

Answer CAS only: There is a very limited connection between the exams and the practice, except for the more advanced CAS exams (mainly Parts 5 [2002], 6 [2002], and 7 [2002]; less so with Parts 8 [2002] and 9 [2002]).

◆◆◆

2.8 | Difficulty of the Examinations

Q Which SOA or CAS examinations did you find to be the most difficult and why? Illustrate your answer with examples.

Answer I can't say I have found one more difficult than others, it is really the time to put into the study of the material that is more difficult. Of course, I found that Course 5 [2002] had much more new material than Courses 1 [2002] through 4 [2002], and it took more time to study for it, which I think is normal. Exam 4 [2002] was maybe more difficult to study for in the sense that it is a mixture of many small parts of material not really related to one another.

Answer Exams 4 [2002] and 6 [2002] were most difficult for me. Exam 4 [2002] requires a lot of memorization of formulas. It is the exam with the least useful content for someone working in my area (pensions). I also found

Exam 6 [2002] difficult, probably mostly because it was my first written exam. The transition from multiple-choice exams to written exams is a difficult one. Written exams require radically different studying and test writing techniques.

Answer   I've tried Exams 1 [2002], 2 [2002], 3 [2002], and 4 [2002]. So far, the fourth one was the most difficult. The amount of formulas to learn by heart is overwhelming. You have to learn a lot of details to pass. And all the material was on subjects that I don't use at work.

Answer   I struggled with SOA Course 6 [2002], because I had very little background in finance. Therefore, I needed to put in a lot of hours to make sure I understood the concepts. Course 8 [2002] was the hardest to date, because it involved less material for which you can study, and more experience-based concepts. For example, the exam contained questions with actual situations a consultant would have to deal with in day-to-day situations (i.e., the client sends an age service table and needs the actuary to calculate the cost of a benefit upgrade). These types of questions were not explicitly dealt with in the syllabus.

Answer   Between the first three, I think that the hardest was the third one. Mostly because there is much more material in this one than in the previous one. Even if I studied about 250 hours, I didn't have time to learn perfectly about topics covered in that exam. However, I still don't know my result, I cannot tell if it was really the hardest one until now. Also, I think the first one is hard but for different reasons. Since it is the first one, you have to learn to do calculation and to answer really quickly. Many personal skills need to be developed, for example, a good way to manage your stress.

Answer   Course 6 [2002], because it is based on advanced finance principles that I don't use at work. My only background was finance courses at university.

Answer   Cannot answer this question from personal experience (only one written!) but I tend to believe that everyone has their own weakness and therefore everyone will find different exams more difficult than another.

Answer   They're all hard. The written answer exams were tough because of the amount of material that needed to be learned. I needed four months of solid studying to pass those. The mathematics exams didn't require as much, but it was different studying—old questions mostly. I only needed two to three months for those. Course 3 [2002] is difficult because it's got some really tough concepts to learn—and memorization doesn't work. You need to understand what is going on.

Answer Course 3 [2002] requires a lot of practice. Course 5 [2002] requires you to spend a lot of time learning lists by heart.

Answer Course 2 [2002] was the most difficult. The questions were not computational, but more theoretical. And even if I did know the answer to a question, it was not obvious which was the correct answer since many of them seemed correct.

Answer Course 8 [2002] (Life). The longest, the most material, the toughest conceptually. The exam included a 17-point question—very hard.

Answer I failed the theory of interest exam because I didn't memorize enough formulas, and ran out of time trying to develop everything from first principles. Once I memorized everything there was no problem.

Answer Part 7C [2002], since it was the first one with Canadian content. I found it difficult because it was my first complete exam (the first five were partitioned when I wrote them) and also because I was still in university and had no experience in reserving and accounting.

Answer Course 5 [2002] for me was the most difficult because it is a very general exam that spans all practices: group insurance, casualty insurance, pensions, life insurance, etc. Since most people have worked only in one field, it is hard to grasp the concepts for all the other fields at once. There's almost too much information to know all at once. You find that you know the section in which you work quite well, but know nothing about the other sections. From a passing perspective, this also means that you will answer your questions quite well on the exam, while others will answer their section's questions quite well, leading all questions to have been very well answered by the people in those sections. This raises that passing bar compared to exams where everyone is in the same boat for all of the questions, and some questions may never be answered well by anybody.

Answer The most difficult, for me, was CAS Course 7 [2002], which deals with annual statement, taxation, and regulation. The difficulty stems from the tremendous amount of minutiae one has to memorize. The exam, however, is important, especially for corporate actuaries who may have to deal with all of the above issues. Taxation in the projection of financial statements. Annual statement understanding is also required for monitoring of company and competitors' results.

Answer Course 8 [2002]. It is the longest exam (now six and a half hours long) and is track-specific. I found it particularly hard because at the time I wrote it, I had less than one year of experience. Therefore, the material included in the exam was all new to me. Moreover, being a Canadian,

I am not familiar with the material used in the United States. There is a lot of U.S. content in Course 8 [2002], and I believe it represents an additional difficulty for Canadian students.

Answer The old CAS Part 8 [2002] (whose material is now partly covered under Part 7 [2002] was definitely the toughest exam I ever wrote. It took me four attempts (this is 4 years!!!) to succeed. This exam was extremely theoretical and very boring. It dealt with "law and insurance," "regulations of insurance," and "government insurance plans." There was more to it, but I do not remember everything. Most of the material was specific to the United States, rendering its learning very painful and rather useless. We had to read and memorize "court cases" (the actual court cases transcripts) (we all wondered what benefit there was behind learning these). Overall, most people I knew felt that learning this material was not the best use of our time. Fortunately, when the new exam system was implemented and Part 8C was torn to pieces, the CAS only kept the most relevant pieces and included them with the new Part 7 [2002]. To sum things up, I had a hard time passing the exam because of the following: (1) the amount of material to learn (about 2000 pages from which questions were picked); and (2) my lack of interest in the topics covered because they were not relevant to Canada and also not relevant in general.

Answer CAS Part 7C [2002] because it involved memorization and no mathematical or numerical problems.

◆◆◆

| 2.9 | **Ways to Pass Examinations** |

Q What were your study tricks and study processes that helped you pass your actuarial examinations? Illustrate your answer with examples related to the SOA and CAS examinations.

Answer Read the books twice, make my own summaries based on my readings and also the summaries available on the market. Read a lot and ask questions to fellow workers who have been there before.

Answer For the multiple-choice exams, do as many practice tests as you can get your hands on. I know of a lot of people who have made the mistake of not getting around to doing the practice tests, because they haven't finished the readings. The practice tests are the most important things to read, try, reread and review. If you can pass the practice tests, you will almost certainly pass the real test. The written answer tests require different preparation techniques. Many people find the first written answer test they write is a real adjustment. The best piece of advice I can give is to memorize lists of important items: even if a direct regurgitation question is not asked, knowing these lists will help you think of the important topics to cover while under pressure.

Answer Do a lot of exercises. I'm not good with just memorizing, I need to practice. So I've done as many exercises as I could. I also tried to vary the sources (not just from the ACTEX [study aids], for example). Start in advance so you don't feel rushed at they end and panic. Set a goal (for example, study 300 hours overall), do a schedule, and note your progress. One week before the exam, go through the books and write a summary sheet with the things you still struggle with.

Answer You need to find a partner that you will not necessarily study together, or be writing the same exam, but someone who will motivate you, and you will push each other to study, and to not procrastinate. At two months before the exam, it is easy to slow down your studying but this is the time that if you have someone else who is also studying, they will not let you relax and take a break!!!

Answer For the first four SOA exams, I made sure that I understood every practice question that I came across. Practice, practice, practice. Course 5 [2002] was a crash course in memorization. I just wrote and rewrote the study notes as much as possible until it sank in! Course 6 was a combination of learning financial and mathematical concepts and some memorization (although the understanding component is much more important). Course 7 [2002] was straightforward: show up at the seminar, follow instructions, and write an essay.

Answer I take a lot of time to study theory and to do the practice exercises of the ACTEX [study aids]. When I finish it, I do sample examinations from previous years in real time. That way, I can have a real idea of my studying status. I think that the more secure way to pass an exam is to study for many hours. For me, there are no other tricks.

Answer I used a lot of memory tricks. For example, in order to remember a whole list of items, I created a word with the first letter of each item of the list, etc. It also really helps to try to apply the concepts we learned (it is easier to remember when you understand the applications).

Answer Start early. In the case of a student going to school and writing exams, starting early is crucial if you do not want to be behind in either. For Course 1 [2002], other than starting early, going through the theory and practicing problems is very important. Doing the same problem over and over again is useful. And a couple of weeks before the exam, doing the sample exams found on the SOA website is extremely useful to familiarize oneself to the format and type of questions which you will be asked. I have found the same type of questions always come up and so understanding the sample exams is very important.

Answer   Here is what worked for me:

- *Practice Exams.* On early exams, do practice exams under timed conditions, learn the rest of the material from an answer guide, and recognize what formulas have to be memorized.
- *Ask Others.* On later written exams, early failure led me to interview successful students on their study techniques, and select some that might work for me. My goal was to pass, not to learn the material better than anyone else.
- *Prepare Early.* Personally, I started early (January/July); a week after the results came out. My goal was one hour per day plus study time—I had a calendar at work where I couldn't ignore it, and put stickers on every day that I met my goal. It is especially important to motivate yourself with short-term goals early in the study schedule. When someone else at work was writing the same exam, I used their progress to pace myself through the material, although I never studied with anyone else. I took a local two-day course for CAS Part 5 [2002] (ratemaking), which scared me into studying even harder for the last month. Otherwise, I don't think much of courses—you can't depend on them to prepare you. I kept asking myself whether what I was doing was going to help me on the day of the exam, and stopped doing it if it wasn't (e.g., writing pretty notes, spending too much time on a problem I couldn't solve), and reminded myself that nothing else (e.g., how much time I had spent) counted when they marked the exam.
- *Make Notes.* I read the material on each paper while writing notes for anything I didn't think I would remember on the day of the exam (this helped me make sure I was absorbing the material, and not just skimming). I concentrated on writing lists and formulas, since it can't be on the exam if they can't make it into a question. I did questions from old exams following each paper to make sure I had learned the right material. In the final weeks before the exam, I reviewed my notes, forming the material into questions on the back of the prior page, so I could study by covering the answers and asking myself the questions. I also did timed exams from prior years. I also think I wrote a good paper—I can read and write quickly, and English is my first language so these are advantages for me. It is important to attempt every question, and for me, not to

go back to the multiple-choice questions as they were very difficult, and if I didn't know the answer the first time, it wasn't going to come to me.

Answer I used homemade flashcards for memorizing lists and concepts for Courses 5 [2002], 6 [2002], and 8 [2002]. Courses 1 [2002] through 4 [2002] were just doing old exam questions over and over again. And I always tried to start studying early (sometimes four to five months in advance).

Answer Mathematics exams (Courses 1 [2002] through 4 [2002]): Do as many practice problems as possible. Classify problems by type for which there is a specific trick to use. Other exams (Courses 5 [2002] through 8 [2002]): Read all material quickly (for background). Spend at least 150 to 200 hours learning lists by heart.

Answer I was always studying with a friend. In my case, learning just by reading was quite difficult. Being able to ask questions to a friend and to go through material with someone else was quite helpful. I recommend it a lot to auditory persons, i.e., those who learn more easily by listening than by reading. Also it helps motivation. We often sat in different rooms and got together to review material after a couple of hours.

Answer Not to think about how stupid it was to make me learn by heart stuff that I knew I would never use again in my life.

Answer I learn lists by acronyms, usually using the first letter of the first words or the first letter of the most relevant word in the sentence. I also remember the number of elements in a list as well as the elements themselves. This way, you don't waste time trying to find the sixth element when there were five items in the list. Do not focus too much on the first reading—you will retain almost nothing of it. I use it only to make notes in my ACTEX manuals when there is not sufficient information for me to understand what is going on. If you don't understand something at first, skip it. Don't lose too much time. Review the material as many times as you can, going into further details each time.

Answer For the mathematical exams I memorized formulae. For the later exams I read the material through once, then went back over it with the ACTEX study guides and prepared index cards with a question on one side and the answer on the other. Once I had been through the entire material again this way I studied exclusively from cards, discarding a card once it was memorized. The cards were a convenient size, allowing me to use all my time on the subway for studying as well. I took my study time in two-hour pieces during the working day.

Answer Here is my advice:

- *Objectives.* Set objectives based on study material (numbers of papers) instead on numbers of hours.
- *Three Readings.* Have at least three readings: the first would be really fast to have a feeling of the material (1 to 2 weeks maximum), the second would be the *real* one where I would read, take notes, and do questions, and the third one would also be fast but for review only.
- *Review Week.* Keep one week to review notes and do past exams. When doing past exams, put yourself in a real situation (3 to 4 hours) to be able to perform similarly at the actual exam.
- *Psychology.* Work on your psychological training. As I was saying before, actuarial exams are similar to sportive competitions. You need to visualize the exam, your performance, and your success.

Answer Starting the study process at least three months in advance. Cramming doesn't work for these puppies. I would sometimes block off a weekend here or there for fun only and take a break, say after the first reading so as to not feel like I was constantly studying. Going through the material several times—using the study aids, using mnemonic devices such as lists of items to be memorized and making a word or phrase out of the first letter of each item on the list to make it easier to remember, participating in study groups or seminars.

Answer It is always important to remember that you are not writing to achieve a certain grade, but to have a better grade than 60% to 65% of the other people who wrote the exam. Unfortunately, we are now caught in a vicious circle, where a fair proportion of candidates failed the exam on the first try. This means that they have a much better understanding of the exam material the second time around. This makes it that much harder to pass for those writing for the first time. Actuarial students need to sit down and figure out their priorities. Is it family? Is it career? Is it passing exams? Unless you clearly want to pass an exam on first sitting each and every time, don't kill yourself studying. Personally, my study method, which seemed to work for me, was:

- *Highlights.* To read through the material once while highlighting the important material (i.e., material which is likely to generate questions). I would read through the highlight a second time while writing a summary (the act of writing it out seemed to help on memorization).
- *Exercises.* Next I would start doing exercises (which helps understanding), while reviewing my notes on an occasional basis. I would always skip the most recent three exams' questions.

- *Old Exams.* A few weeks before the exam, I would write the exam from three years ago under exam conditions. This is usually a very good wake-up call, to see that one is nowhere near ready. I would do the same one week later, and a few days before the exam.
- *Priorities.* If one is dedicated to the exam process, one should not go out on Friday or Saturday nights. That's why it is so important to set up your priorities. I would always do my own summary, never relying on the ACTEX summary since I found them to be, in some instances, inaccurate. When reading through the material, always keep in mind: Is this something I would ask a question on? The ability to anticipate the material for questions is a great help in focusing the study hours on important subjects.

Answer This is what I would do:

- *Two Readings.* Read the material twice.
- *Notes.* Take summary notes, including the creation of tables to sort out similar concepts.
- *Drill.* Practice with problems a lot.
- *Last Month.* For more advanced exams, spend about a month at the end of the study period to learn all my personal notes (in some cases, I had more than 300 pages of notes).

Answer In studying for actuarial exams, what helped me most was planning and discipline. Planning is important because of the quantity of work involved:

- *Tables.* I start by creating a table listing all readings from the syllabus, to which I add columns indicating whether I've read the article, typed the notes, worked the problems (twice), and reviewed the article.
- *Planning.* Based on the number of articles and pages to read, I determine how many hours I need to study. I usually plan 400 hours per exam. Once the planning stage is completed, discipline is required to stick to the plan.
- *Order.* I start by reading the articles in the order they are presented in the syllabus (they are already organized by topic).
- *Summaries.* After reading an article, I type notes summarizing important concepts and definitions, and include mathematical examples.
- *Practice.* I then work the practice problems from the ACTEX manual.
- *Procedure.* I repeat the same procedure for each article. It usually takes two months to read, type the notes, and go through the first round of problems.

- *One Reading.* Because my notes are very thorough, I read the articles only once.
- *Review.* After that, I start reviewing my notes and working the problems for a second time, which takes about a month.
- *Memorize.* Two weeks before the exam, I start memorizing lists, definitions, and formulas.
- *Sample Exam.* Two days before the exam, I take a few practice exams.
- *Timing.* Every day, I record my study time so I can see my progress. At the beginning of each week, I determine how many hours I need to study and how many articles I need to complete to be on schedule. By doing so, what seems to be a mountain of work is broken down into more manageable pieces.

♦♦♦

2.10 | Study Aids

Actuarial examinations are difficult to pass. The average pass rate is usually below 40%. This means, of course, that the average failure rate is often more than 60%. In other words, many bright and dedicated students who are used to getting high grades in college are having to adjust to the fact that they may actually fail an examination.

To help increase the students' chances of success in actuarial examinations, an extensive commercial support system has developed. Various companies market different types of study tools, organize seminars and special courses, and provide other help for a fee. The survey upon which this book is based suggests that the following study tools are often used:

- The ACTEX study material, produced by ACTEX Publications, Mad River Books, and ACTEX Actuarial Recruiting.
- The ASM study material, also marketed by ACTEX.
- The JAM study material, also marketed by ACTEX.
- The How-To-Pass study manuals, marketed by How-To-Pass. (These manuals are now available only from second-hand sources.)
- The Study Aids, marketed by NEAS [New England Actuaries Seminars].

The list is not exhaustive. A search on the Internet will produce additional tools not mentioned by the respondents to the survey. Among these are the study tools made available by CAS on its website. The material includes previous examinations, an exam study group, and pass/fail statistics.

Q Which study aids, such as ACTEX, have you used and which would you recommend? Illustrate your answer with examples related to the SOA and CAS examinations.

Answer Examinations 1 [2002] through 4 [2002]: How-To-Pass were great! Exam 5 [2002]: JAM was great! I have found that ACTEX have too much information and not enough explanations. How-To-Pass and JAM really *talk to you*, they are not only summaries of concepts and formulas.

Answer Only ACTEX, as the others didn't exist at the time.

Answer I have actually tried out summaries from all four providers. In my mind, ACTEX is the best all-round, but the ASM books were also really good. I think it is often a good idea to get books from two different companies for a given test, as it is amazing how different the content will be between the various summaries (especially for the higher level tests). The tests are not set by the people who write the study guides, and it is up to the candidates to "guess" what topics and questions will actually be on an SOA test. I have found that most study guides do not adequately cover certain topics, and that using two different study guides offer a certain level of assurance that nothing will be "missed." The textbooks, on the other hand, generally range from mostly useless to completely useless. If time is an issue, and something has to be cut from your study plan, start with the textbooks.

Answer I've used the ACTEX for every exam. It helps, but I often find that they do not explain very well. They take for granted that you know a lot of things. I used it mostly for the exercises. I also use the flashcards. I find them handy because you can always have them on you (in the bus, waiting in line, etc.) and they help me memorize the formulas. I've used the How-To-Pass. I thought they were skipping too much material. They were basically advising us to memorize only a couple formulas. But we don't have time to derive formulas during the exam, and knowing a couple more formulas can make a difference between a 5 and a 6. After all this, I use Study Aids. So far, I like them best. It's clear, they don't explain too fast and there are a lot of exercises. Sometimes, exercises are repetitive, but I remember better by repetition.

Answer For Courses 1 [2002] through 3 [2002], I used the ACTEX manual, which provides very good summaries, and covers all the topics needed to pass the exam.

Answer For the first four SOA exams, I used the ACTEX manual almost exclusively. Combined with previous exams, I just kept doing practice problems. For Course 5 [2002] I used JAM to memorize the material. For Course 6, I also used JAM to memorize, but the actual books were definitely very important for understanding the concepts. I've never used ASM.

Answer The only books I have used were ACTEX manuals.

Answer I only used the ACTEX manuals and really liked them.

Answer For Course 1 [2002], I personally felt that the ACTEX manual did not help at all. The questions are too basic and do not illustrate the actual course. The manual is useful if someone does not know the basics, but to prepare for an SOA exam, I did not find it helpful. For the later courses, I have heard the same type of comments about the ACTEX manual.

Answer I used the ACTEX manuals. The other manuals weren't available when I took exams, for the most part. I never went to exam study seminars that cost a ton of money. Students in the United States seem to like them, but I thought they were too expensive for what you got. The new *8I Made Easy* manual, available as a free Internet download, was also helpful.

Answer JAM provides a better summary than ACTEX. For mathematics exams, ACTEX may have more practice problems.

Answer ACTEX is quite good. I do not know the others. I have used the ACTEX manuals for all my exams.

Answer Old exams were my best tools. ACTEX manuals were used a little bit as well.

Answer Using study guides could be good. Do not overemphasize them. The most important thing to really study are old sample SOA exams. The questions come back from year to year. It is possible to learn to answer all the questions without knowing them perfectly. Actually one does not imply the other. In my experience, Exams 1 through 3 have almost no original questions, i.e., questions which have not appeared on a previous exam.

Answer I have always used the ACTEX manuals and they worked for me. Whichever tool you use, I recommended using only one for a given examination. As I remember things visually (by the disposition on the page), seeing it in two or three different ways is more confusing than helpful. Should you be short on time, use JAM (it skips corners). If you want to be more thorough, use ACTEX.

Answer ACTEX. I used them in making index cards. I have never used the others.

Answer The only one that I used was ACTEX. At my time, it was really complete and was really helpful.

Answer I've used the JAM and ASM. I would recommend these over ACTEX, if available because these are usually written by people who take the time to go through the material and make an intelligent comment about the syllabus as opposed to ACTEX, which is just a regurgitation of the textbooks with loads and loads of mistakes and typos, and often never peer reviewed before printing. JAM for Course 5 [2002] is my favorite. SOA Course 5 is evil, but using the JAM makes it a bit easier. You can order the cue cards. They are excellent for review.

Answer I always used ACTEX to help me with my study. The important word is help. I found that ACTEX is not always accurate, and cannot replace proper reading and understanding of the study material. ACTEX, in my mind, is most useful as a compilation of old exam questions.

Answer I have used ACTEX for SOA exams and Casualty Study Manuals for CAS exams.

Answer The ACTEX manuals provide a good sample of past questions. However, I don't use their summaries. I prefer to type my own notes.

The study material quoted in the survey, with the exception of out-of-print "How-to-Pass" manuals, can be found on the following websites:

- ACTEX: www.actexmadriver.com
- ASM: www.studymanuals.com
- CAS: www.casact.org/admissions/studytools
- JAM: www.studyjam.com
- Study Aids: www.neas-seminars.com/misc

Although the survey gives preference to one or two of the six listed study tools, the list of mentioned tools is far from complete. Depending on the search engine used, the Internet query *actuarial study tools*, for example, can produce over 20,000 hits. Since material posted on the Internet is often changed and updated without notice, the mentioned references may have been modified. The examples discussed here and many more like them can be found on the website of the Society of Actuaries. At each stage, the posted examinations and solutions cover at least the ten last years.

◆◆◆

2.11 | Mathematical Foundations of Actuarial Science

Ideas and Techniques

The material in this course dealt with the mathematical foundations of actuarial science. According to the Society of Actuaries, *the purpose of this course is to develop a knowledge of the fundamental mathematical tools for quantitatively assessing risk. The application of these tools to problems encountered in actuarial science is emphasized. A thorough command of calculus and probability topics is assumed. Additionally, a very basic knowledge of insurance and risk management is assumed.* (See [3], Page 23.)

Why does an actuary need to know calculus? You probably know that some of the basic objects of calculus are the real numbers and differentiable and integrable functions on the real numbers.

Where are real numbers needed in actuarial science? Let us consider a very basic example. Among other topics, actuarial science deals with the cost of money over time. Interest is the cost of money. You pay interest when you borrow money and you earn interest when you lend money. The amount of interest involved is a function of the amount borrowed, the rate of interest charged, and the length of time for which the money is borrowed. Moreover, in most cases, the interest charged or earned is compound interest. It is calculated over shorter periods than the entire lending period.

Many of the mathematical ideas upon which actuarial science is based are hundreds of years old and have stood the test of time. Calculus dates back to Newton (1642–1727) and Leibniz (1646–1716). History tells us that the theory of probability even predates the beginning of calculus. It is said that a professional gambler named Chevalier de Mere made a great deal of money by betting people that by rolling a die four times he could get at least one six. He was so successful at it that he soon had trouble finding people willing to play his game. So he changed the rules. He started to bet that he could get at least two sixes by rolling a die twenty-four times. Unfortunately for him, he systematically lost. He contacted Pascal (1623–1662) to help explain his losses. Pascal began to correspond with Fermat (1601–1665) to analyze the problem and it is said that thus probability theory was born. Course 1 [2002] builds on the ideas and techniques of calculus and probability.

Example 1 *Continuous interest*

Suppose you borrow P dollars at x percent interest for n periods. Using geometric progressions, we can develop a formula for the cost of the loan:

$$P(1+x)^n$$

The formula is based on the assumption that the interest is calculated at the end of each period. However, if the interest is compounded at intervals that are shorter than the interest rate period, then the cost of the loan becomes

$$P\left(1+\frac{x}{t}\right)^{tn}$$

and since

$$(1+x) < \left(1+\frac{x}{2}\right)^2 < \left(1+\frac{x}{3}\right)^3 < \cdots$$

we see that compounding over shorter periods increases the interest paid on the loan. How far can we increase this compounding factor? It turns out that there is a limit beyond which we cannot go. It is called *continuous interest*. Since

$$\lim_{t\to\infty}\left(1+\frac{x}{t}\right)^t = e^x$$

the largest amount of interest that can be charged for borrowing P dollars at x percent interest for n periods by increasing the compounding intervals to their limit is

$$Pe^{xn}$$

As we can see, even at this very elementary level of finance, the number e, one of the most celebrated of the real numbers, plays a pivotal role. So do limits, the exponential function e^x, and the idea of continuity. We need calculus to understand these concepts. ▲

It is of course important that these ideas enter the actuarial world at its foundation. Calculus usually remains in the background in day-to-day work of an actuary.

Examination Topics

The examination consisted of forty multiple-choice questions. They dealt with the following topics from the SOA and CAS syllabus:

Q1 Investment models, exponential functions, logarithmic functions.

Q2 Stocks, dividends, geometric progressions, logarithmic functions.

Q3 Parametric curves, velocity vectors, lengths of vectors, derivatives, cosine functions.

Q4 Random variables, independent random variables, distribution functions, density functions, expected values, maximum-value functions, probabilities, sine and cosine functions, integrals.

Q5 Functions defined by cases, property and casualty insurance, density functions, joint density functions, probabilities, double integrals.

Q6 Life insurance (standard, preferred, and ultra-preferred), probabilities, conditional probabilities, Bayes' formula.

Q7 Functions defined by cases, joint density functions, covariance, double integrals, expected values.

Q8 Capital, labor, production rates of change, chain rule.

Q9 Health insurance, risk factors, probabilities, unconditional probabilities, algebra of sets.

Q10 Life insurance, premiums, survival functions, expected values, probabilities.

Q11 Insurance products, functional models, derivatives, maximum-value test.

Q12 Probabilities, algebra of sets.

Q13 Health insurance, probabilities, algebra of sets, independence of events.

Q14 Functions defined by cases, stock prices modeled with random variables, joint density distributions, conditional variance, marginal density functions, integrals, expected values.

Q15 Parametric curves, slopes, tangents, derivatives.

Q16 Functions defined by cases, ratios, graphs, concavity, step functions.

Q17 Functions defined by cases, exponential functions, auto insurance, probability distributions, probability density functions, expected values, integrals, maximum-value test.

Q18 Exponential functions, partial derivatives, rates of change.

Q19 Lifetimes, independent lifetimes, means, variances, normal distributions, random variables, probabilities, minimum values.

Q20 Continuous measurements, time-to-failure, exponential distributions, means, expected values, integrals, maximum values.

Q21 Differential equation models for diseases, separable differential equations, general solutions, partial fractions, integrals.

Q22 Auto insurance, insurance claims, random variables, exponential distributions, means, probabilities, algebra of sets.

Q23 Quality control, probabilities, Bayes' formula.

Q24 Lifetimes, joint density functions, probabilities, double integrals.

Q25 Volumes, surface areas, spheres, rates of change, derivatives.

Q26 Earthquake insurance, premiums modeled by random variables, exponential random variables, independent random variables, probability density functions, means, double integrals, derivatives.

Q27 Functions defined by cases, property and casualty insurance, damage claims, independent random variables, joint density functions, expected values, derivatives, improper integrals.

Q28 Functions defined by cases, mortality functions integrals, inequalities, minimum values.

Q29 Insurance risk modeled by random variables, probabilities, trinomials, probability functions.

Q30 Profit models, fixed costs, variable costs, maximum profit, cost functions, revenue functions, profit functions, quadratic functions.

Q31 Group health insurance, supplementary coverage, probabilities, Venn diagrams.

Q32 Time-to-failure, exponential distributions, means, variances.

Q33 Insurance claims, normal distribution of claims, means, standard deviations, independent random variables, expected values, linear combinations.

Q34 Graphs of functions, graphs of derivatives, slopes.

Q35 Property and casualty insurance, time-to-failure, density functions, variances, expected values, integrals.

Q36 Pollution models, averages, double integrals.

Q37 Loss models, probabilities, expected values.

Q38 Functions defined by cases, continuity at a point.

Q39 Functions defined by cases, home insurance, random variables, probability distributions, density functions, integrals.

Q40 Life insurance, probabilities, conditional probabilities, algebra of sets.

If we examine the frequencies of some of the topics and techniques tested in this examination, we come up with the following result: probability (30/40), functions (20/40), integrals (15/40), random variables (14/40), derivatives (10/40), expected values (10/40).

Examination Questions and Answers

Here are some examples of how these ideas and techniques were tested.

Question 1 *A company agrees to accept the highest of four sealed bids on a property. The four bids are regarded as four independent random variables with common cumulative distribution function*

$$F(x) = \frac{1}{2}(1 + \sin \pi x)$$

for $\frac{3}{2} \leq x \leq \frac{5}{2}$. What is the expected value of the accepted bid?

Answer Let X_1, X_2, X_3, and X_4 denote the four independent bids with common distribution function F. Then if we define $Y = \max(X_1, X_2, X_3, X_4)$, the distribution function G of Y is given by

$$
\begin{aligned}
G(y) &= \Pr[Y \leq y] \\
&= \Pr[(X_1 \leq y) \cap (X_2 \leq y) \cap (X_3 \leq y) \cap (X_4 \leq y)] \\
&= \Pr[X_1 \leq y] \Pr[X_2 \leq y] \Pr[X_3 \leq y] \Pr[X_4 \leq y] \\
&= [F(y)]^4 \\
&= \frac{1}{16}(1 + \sin \pi y)^4, \quad \frac{3}{2} \leq y \leq \frac{5}{2}
\end{aligned}
$$

It follows that the density function g of Y is given by

$$
\begin{aligned}
g(y) &= G'(y) \\
&= \frac{1}{4}(1 + \sin \pi y)^3 (\pi \cos \pi y) \\
&= \frac{\pi}{4} \cos \pi y (1 + \sin \pi y)^3, \quad \frac{3}{2} \leq y \leq \frac{5}{2}
\end{aligned}
$$

Therefore,

$$
\begin{aligned}
E[Y] &= \int_{3/2}^{5/2} y g(y) \, dy \\
&= \int_{3/2}^{5/2} \frac{\pi}{4} y \cos \pi y (1 + \sin \pi y)^3 \, dy \\
&\approx 2.22656
\end{aligned}
$$

This answers the question. ▲

Question 2 *Two life insurance policies, each with a death benefit of 10,000 and a one-time premium of 500, are sold to a couple, one for each person. The policies will expire at the end of the tenth year. The probability that only the wife will survive at least ten years is 0.025, the probability that only the husband will survive at least ten years is 0.01, and the probability that both of them will survive at least ten years is 0.96. What is the expected excess of premiums over claims, given that the husband survives at least ten years?*

Answer Let W be the event that the wife survives at least 10 years, H the event that the husband survives at least 10 years, B the paid benefit, and P the profit from selling the policies. Then

$$\Pr[H] = \Pr[H \cap W] + \Pr[H \cap W^c] = 0.96 + 0.01 = 0.97$$

and

$$\Pr[W^c \mid H] = \frac{\Pr[H \cap W^c]}{\Pr[H]} = \frac{0.01}{0.97} = 0.0103$$

It follows that

$$\begin{aligned}
E[P] &= E[1000 - B] \\
&= 1000 - E[B] \\
&= 1000 - \{(0)\Pr[W \mid H] + (10{,}000)\Pr[W^c \mid H]\} \\
&= 1000 - 10{,}000\,(0.0103) \\
&= 1000 - 103 \\
&= 897
\end{aligned}$$

This answers the question. ▲

Question 3 *An insurance company has 160,000 to spend on the development and marketing of a new insurance policy. If x is spent on development and y is spent on marketing, then*

$$\frac{x^{1/4}y^{3/4}}{1000}$$

policies will be sold during the first year. Calculate the maximum possible number of policies the company can sell during the first year.

Answer Observe that x and y follow the constraint equation

$$x + y = 160{,}000$$
$$x = 160{,}000 - y, \text{ where } 0 \le y \le 160{,}000$$

Using this constraint equation, we express the policy sales $g(x,y)$ as a function $f(y)$ of marketing y:

$$f(y) = g(160{,}000 - y, y) = 0.001\,(160{,}000 - y)^{1/4}\,y^{3/4}$$

We then compute $f'(y)$:

$$f'(y) = \left\{ -\frac{1}{4}(160,000-y)^{-3/4}y^{3/4} + \frac{3}{4}(160,000-y)^{1/4}y^{-1/4} \right\} / 1000$$

$$= -\frac{1}{4000}(160,000-y)^{-3/4}y^{-1/4}[y - 3(160,000-y)]$$

$$= -\frac{1}{4000}(160,000-y)^{-3/4}y^{-1/4}(4y - 480,000)$$

$$= \frac{1}{1000}(160,000-y)^{-3/4}y^{-1/4}(120,000-y), \quad 0 \le y \le 160,000$$

and note that

$$f'(y) > 0 \text{ for } 0 \le y < 120,000$$
$$f'(y) = 0 \text{ for } y = 120,000$$
$$f'(y) < 0 \text{ for } 120,000 < y < 160,000$$

Therefore sales are maximized when $y = 120,000$. It follows that

$$f(120,000) = 0.001(160,000 - 120,000)^{1/4}(120,000)^{3/4} = 91.2$$

maximizes f. ▲

Question 4 *A study is being conducted in which the health of two independent groups of ten policyholders is being monitored over a one-year period of time. Individual participants in the study drop out before the end of the study with probability 0.2 (independently of the other participants). What is the probability that at least nine participants complete the study in one of the two groups, but not in both groups?*

Answer Let X be the number of group 1 participants that complete the study, and let Y be the number of group 2 participants that complete the study. Since

$$P[X \ge 9] = \left[\binom{10}{9}(0.2)(0.8)^9 + \binom{10}{10}(0.8)^{10} \right] = 0.376,$$

it follows that

$$P\{[(X \ge 9) \cap (Y < 9)] \cup [(X < 9) \cap (Y \ge 9)]\}$$
$$= P[(X \ge 9) \cap (Y < 9)] + P[(X < 9) \cap (Y \ge 9)]$$
$$= 2P[(X \ge 9) \cap (Y < 9)] \text{ (due to symmetry)}$$
$$= 2P[X \ge 9]P[Y < 9] \text{ (due to independence)}$$
$$= 2P[X \ge 9]P[X < 9] \text{ (again due to symmetry)}$$
$$= 2P[X \ge 9](1 - P[X \ge 9])$$
$$= 2[0.376][1 - 0.376] = 0.469$$

This answers the question. ▲

Question 5 *The stock prices of two companies at the end of any given year are modeled with random variables X and Y that follow a distribution with joint density function*

$$f(x,y) = \begin{cases} 2x & \text{for } 0 < x < 1, \ x < y < x+1 \\ 0 & \text{otherwise} \end{cases}$$

What is the conditional variance of Y given that X = x?

Answer Let $f_1(x)$ denote the marginal density function of X. Then

$$f_1(x) = \int_x^{x+1} 2x \, dy = 2xy \Big|_x^{x+1} = 2x(x+1-x) = 2x, \ 0 < x < 1$$

Consequently,

$$f(y \mid x) = \frac{f(x,y)}{f_1(x)} = \begin{cases} 1 & \text{if } x < y < x+1 \\ 0 & \text{otherwise} \end{cases}$$

and

$$E[Y \mid X] = \int_x^{x+1} y \, dy = \frac{1}{2} y^2 \Big|_x^{x+1} = \frac{1}{2}(x+1)^2 - \frac{1}{2}x^2$$

$$= \frac{1}{2}x^2 + x + \frac{1}{2} - \frac{1}{2}x^2 = x + \frac{1}{2}$$

$$E[Y^2 \mid X] = \int_x^{x+1} y^2 \, dy = \frac{1}{3} y^3 \Big|_x^{x+1} = \frac{1}{3}(x+1)^3 - \frac{1}{3}x^3$$

$$= \frac{1}{3}x^3 + x^2 + x + \frac{1}{3} - \frac{1}{3}x^3 = x^2 + x + \frac{1}{3}$$

$$\text{Var}[Y \mid X] = E[Y^2 \mid X] - \{E[Y \mid X]\}^2 = x^2 + x + \frac{1}{3} - \left(x + \frac{1}{2}\right)^2$$

$$= x^2 + x + \frac{1}{3} - x^2 - x - \frac{1}{4} = \frac{1}{12}$$

This answers the question. ▲

Question 6 *An auto insurance company insures an automobile worth 15,000 for one year under a policy with a 1,000 deductible. During the policy year there is a 0.04 chance of partial damage to the car and a 0.02 chance of a total loss of the*

car. If there is partial damage to the car, the amount X of damage (in thousands)
follows a distribution with density function

$$f(x) = \begin{cases} 0.5003e^{-x/2} & \text{for } 0 < x < 15 \\ 0 & \text{otherwise} \end{cases}$$

What is the expected claim payment?

Answer Let Y denote the claim payment made by the insurance company. Then

$$Y = \begin{cases} 0 & \text{with probability } 0.94 \\ \max(0, X - 1) & \text{with probability } 0.04 \\ 14 & \text{with probability } 0.02 \end{cases}$$

and

$$E[Y] = (0.94)(0) + (0.04)(0.5003)\int_1^{15}(x-1)e^{-x/2}dx + (0.02)(14)$$

$$= (0.020012)\left[\int_1^{15}xe^{-x/2}dx - \int_1^{15}e^{-x/2}dx\right] + 0.28$$

$$= 0.28 + (0.020012)\left[-2xe^{-x/2}\Big|_1^{15} + 2\int_1^{15}e^{-x/2}dx - \int_1^{15}e^{-x/2}dx\right]$$

$$= 0.28 + (0.020012)\left[-30e^{-7.5} + 2e^{-0.5} + \int_1^{15}e^{-x/2}dx\right]$$

$$= 0.28 + (0.020012)\left(-30e^{-7.5} + 2e^{-0.5} - 2e^{-7.5} + 2e^{-0.5}\right)$$

$$= 0.28 + (0.020012)\left(-32e^{-7.5} + 4e^{-0.5}\right)$$

$$= 0.28 + (0.020012)(2.408)$$

$$= 0.328 \text{ (in thousands)}$$

It follows that the expected claim payment is 328. ▲

Question 7 *A company manufactures light bulbs with a lifetime, in months, that*
is normally distributed with mean 3 and variance 1. A consumer buys a number
of these bulbs with the intention of replacing them successively as they burn out.

The light bulbs have independent lifetimes. What is the smallest number of bulbs to be purchased so that the succession of light bulbs produces light for at least 40 months with probability at least 0.9772?

Answer Let $X_1, \ldots, X_n$ denote the life spans of the n light bulbs purchased. Since these random variables are independent and normally distributed with mean 3 and variance 1, the random variable

$$S = X_1 + \cdots + X_n$$

is also normally distributed with mean $\mu = 3n$ and standard deviation $\sigma = \sqrt{n}$. We want to choose the smallest value for n such that

$$0.9772 \leq \Pr[S > 40] = \Pr\left[\frac{S - 3n}{\sqrt{n}} > \frac{40 - 3n}{\sqrt{n}}\right]$$

Recalling that in the case of a normal distribution, the probability that an observation is above two standard deviations below the mean is 0.9772, we conclude that n should satisfy the following inequality:

$$-2 \geq \frac{40 - 3n}{\sqrt{n}}$$

To find such an n, we solve the corresponding equation for n:

$$-2 = \frac{40 - 3n}{\sqrt{n}}$$
$$-2\sqrt{n} = 40 - 3n$$
$$3n - 2\sqrt{n} - 40 = 0$$
$$\left(3\sqrt{n} + 10\right)\left(\sqrt{n} - 4\right) = 0$$
$$\sqrt{n} = 4$$
$$n = 16$$

This answers the question. ▲

Question 8 *The rate at which a disease spreads through a town can be modeled by the differential equation*

$$\frac{dQ}{dt} = Q(N - Q)$$

where $Q(t)$ is the number of residents infected at time t and N is the total number of residents. Find $Q(t)$.

Answer The given equation has the obvious singular solutions $Q(t) \equiv 0$ and $Q(t) \equiv N$, but these are of little relevance to the given problems. To find the other solutions, we proceed as follows. The differential equation that we are given is separable. As a result, the general solution is given by

$$\int \frac{1}{Q(N-Q)} dQ = \int dt = t + C$$

where C is a constant. To calculate the integral on the left-hand side of this equation, we determine the partial fractions of the integrand. In other words, we need to find constants A and B such that

$$\frac{1}{Q(N-Q)} = \frac{A}{Q} + \frac{B}{N-Q}$$

$$1 = A(N-Q) + BQ$$

$$1 = AN + (B-A)Q$$

Therefore, $AN = 1$ and $B - A = 0$. Hence $B = A = 1/N$ and

$$\int \frac{1}{Q(N-Q)} dQ = \frac{1}{N} \int \frac{1}{Q} + \frac{1}{N} \int \frac{1}{N-Q} dQ$$

$$= \frac{1}{N} \ln Q - \frac{1}{N} \ln(N-Q) + K$$

$$= \frac{1}{N} \ln \left[\frac{Q}{N-Q} \right] + K$$

where K is a constant.

Consequently,

$$\frac{1}{N} \ln \left[\frac{Q}{N-Q} \right] + K = t + C$$

$$\left(\frac{Q}{N-Q} \right)^{1/N} e^K = e^t e^C$$

$$\left(\frac{Q}{N-Q} \right)^{1/N} = e^t e^{C-K}$$

$$\frac{Q}{N-Q} = e^{Nt} e^{N(C-K)}$$

so that

$$Q = ae^{Nt}(N-Q)$$

$$= aNe^{Nt} - ae^{Nt}Q, \text{ where } a = e^{N(C-K)} \text{ is a constant}$$

$$(1+ae^{Nt}) Q = aNe^{Nt}$$

$$Q(t) = \frac{aNe^{Nt}}{1+ae^{Nt}}$$

This answers the question. ▲

Question 9 *A company offers earthquake insurance. Annual premiums are modeled by an exponential random variable with mean 2. Annual claims are modeled by an exponential random variable with mean 1. Premiums and claims are independent. Let X denote the ratio of claims to premiums. What is the density function of X?*

Answer Let U be the annual claims, V the annual premiums, $g(u,v)$ the joint density function of U and V, $f(x)$ the density function of X, and $F(x)$ the distribution function of X. Then, since U and V are independent,

$$g(u,v) = (e^{-u}) \left(\frac{1}{2} e^{-v/2}\right) = \frac{1}{2} e^{-u} e^{-v/2}, \quad 0 < u < \infty, \ 0 < v < \infty$$

and

$$F(x) = \Pr\left[\frac{U}{V} \le x\right] = \Pr[U \le Vx]$$

$$= \int_0^\infty \int_0^{vx} g(u,v)\,du\,dv = \int_0^\infty \int_0^{vx} e^{-u} e^{-v/2}\,du\,dv$$

$$= \int_0^\infty -\frac{1}{2} e^{-u} e^{-v/2} \Big|_0^{vx} dv = \int_0^\infty \left(-\frac{1}{2} e^{-vx} e^{-v/2} + \frac{1}{2} e^{-v/2}\right) dv$$

$$= \int_0^\infty \left(-\frac{1}{2} e^{-v(x+1/2)} + \frac{1}{2} e^{-v/2}\right) dv$$

$$= \left[\frac{1}{2x+1} e^{-v(x+1/2)} - e^{-v/2}\right]_0^\infty$$

$$= -\frac{1}{2x+1} + 1$$

It follows that

$$f(x) = F'(x) = \frac{2}{(2x+1)^2}$$

This answers the question. ▲

Question 10 *Claim amounts for wind damage to insured homes are independent random variables with common density function*

$$f(x) = \begin{cases} \frac{3}{x^4} & for\ x > 1 \\ 0 & otherwise \end{cases}$$

where x is the amount of a claim in thousands. Suppose three such claims will be made. What is the expected value of the largest of the three claims?

Answer First, observe that the distribution function of X is given by

$$F(x) = \int_1^x \frac{3}{t^4} dt = -\frac{1}{t^3}\Big|_1^x = 1 - \frac{1}{x^3}, \quad x > 1$$

Next, let X_1, X_2, and X_3 denote the three claims made that have this distribution. Then if Y denotes the largest of these three claims, it follows that the distribution function of Y is given by

$$G(y) = \Pr[X_1 \le y] \Pr[X_2 \le y] \Pr[X_3 \le y]$$

$$= \left(1 - \frac{1}{y^3}\right)^3, y > 1$$

while the density function of Y is given by

$$g(y) = G'(y) = 3\left(1 - \frac{1}{y^3}\right)^2 \left(\frac{3}{y^4}\right) = \left(\frac{9}{y^4}\right)\left(1 - \frac{1}{y^3}\right)^2, y > 1$$

Therefore,

$$E[Y] = \int_1^\infty \frac{9}{y^3}\left(1 - \frac{1}{y^3}\right)^2 dy = \int_1^\infty \frac{9}{y^3}\left(1 - \frac{2}{y^3} + \frac{1}{y^6}\right) dy$$

$$= \int_1^\infty \left(\frac{9}{y^3} - \frac{18}{y^6} + \frac{9}{y^9}\right) dy = -\frac{9}{2y^2} + \frac{18}{5y^5} - \frac{9}{8y^8}\Big|_1^\infty$$

$$= 9\left[\frac{1}{2} - \frac{2}{5} + \frac{1}{8}\right] = 2.025 \text{ (in thousands)}$$

This answers the question. ▲

Question 11 *A large pool of adults earning their first driver's license includes 50% low-risk drivers, 30% moderate-risk drivers, and 20% high-risk drivers. Because these drivers have no prior driving record, an insurance company considers each driver to be randomly selected from the pool. This month, the insurance* *company writes four new policies for adults earning their first driver's license. What is the probability that these four will contain at least two more high-risk drivers than low-risk drivers?*

Answer Let X be the number of low-risk drivers insured, Y the number of moderate-risk drivers insured, Z the number of high-risk drivers insured, and $f(x,y,z)$ the probability function of X, Y, and Z. Then f is a trinomial probability function, so

$$\Pr[z \geq x+2] = f(0,0,4) + f(1,0,3) + f(0,1,3) + f(0,2,2)$$

$$= (0.20)^4 + 4(0.50)(0.20)^3$$

$$+ 4(0.30)(0.20)^3 + \frac{4!}{2!2!}(0.30)^2(0.20)^2$$

$$= 0.0488$$

This answers the question. ▲

Question 12 *A town in the shape of a square with each side measuring 4 has an industrial plant at its center. The industrial plant is polluting the air such that the concentration of pollutants at each location (x,y) in the town can be modeled by the function*

$$C(x,y) = 22,500\left(8 - x^2 - y^2\right) \text{ for } -2 \leq x \leq 2 \text{ and } -2 \leq y \leq 2.$$

Calculate the average pollution concentration over the entire town.

Answer Let T denote the total concentration of pollutants over the town. By symmetry we have

$$T = 4 \int_0^2 \int_0^2 22,500\left(8 - x^2 - y^2\right) dx dy$$

$$= (4)(7500) \int_0^2 \left(24x - x^3 - 3xy^2\right)\big|_0^2 dy$$

$$= 30,000 \int_0^2 \left(48 - 8 - 6y^2\right) dy$$

$$= 30,000 \int_0^2 \left(40 - 6y^2\right) dy$$

$$= 30,000 \left(40y - 2y^3\right)\big|_0^2 = 30,000(80 - 16)$$

$$= 30,000(64) = 1,920,000$$

Since the town covers 16 square miles, it follows that the average pollution concentration A is

$$A = T/16 = 1,920,000/16 = 120,000$$

This answers the question. ▲

Question 13 *A tour operator has a bus that can accommodate* 20 *tourists. The operator knows that tourists may not show up, so he sells* 21 *tickets. The probability that an individual tourist will not show up is* 0.02, *independent of all other tourists. Each ticket costs* 50, *and is non-refundable if a tourist fails to show up. If a tourist shows up and a seat is not available, the tour operator has to pay* 100 *(ticket cost + 50 penalty) to the tourist. What is the expected revenue of the tour operator?*

Answer Observe that the bus driver collects $21 \times 50 = 1050$ for the 21 tickets he sells. However, he may be required to refund 100 to one passenger if all 21 ticket holders show up. Since passengers show up or do not show up independently of one another, the probability that all 21 passengers will show up is

$$(1 - 0.02)^{21} = (0.98)^{21} = 0.65$$

Therefore, the tour operator's expected revenue is $1050 - (100)(0.65) = 985$. ▲

Question 14 *An insurance company insures a large number of homes. The insured value X of a randomly selected home is assumed to follow a distribution with density function*

$$f(x) = \begin{cases} 3x^{-4} & \text{for } x > 1 \\ 0 & \text{otherwise} \end{cases}$$

Given that a randomly selected home is insured for at least 1.5, *what is the probability that it is insured for less than* 2?

Answer Let F denote the distribution function of f. Then

$$F(x) = \Pr[X \le x] = \int_1^x 3t^{-4}\,dt = -t^{-3}\big|_1^x = 1 - x^{-3}$$

Therefore,

$$\Pr[X < 2 \mid X \geq 1.5] = \frac{\Pr[(X < 2) \cap (X \geq 1.5)]}{\Pr[X \geq 1.5]}$$

$$= \frac{\Pr[X < 2] - \Pr[X \leq 1.5]}{\Pr[X \geq 1.5]}$$

$$= \frac{F(2) - F(1.5)}{1 - F(1.5)} = \frac{(1.5)^{-3} - (2)^{-3}}{(1.5)^{-3}}$$

$$= 1 - \left(\frac{3}{4}\right)^3 = 0.578$$

This answers the question. ▲

Question 15 *A public health researcher examines the medical records of a group of 937 men who died in 1999 and discovers that 210 of the men died from causes related to heart disease. Moreover, 312 of the 937 men had at least one parent who suffered from heart disease, and, of these 312 men, 102 died from causes related to heart disease. Determine the probability that a man randomly selected from this group died of causes related to heart disease, given that neither of his parents suffered from heart disease.*

Answer Let H be the event that a death is due to heart disease, and F be the event that at least one parent suffered from heart disease.

Based on the medical records, we have

$$P[H \cap F^c] = \frac{210 - 102}{937} = \frac{108}{937}$$

$$P[F^c] = \frac{937 - 312}{937} = \frac{625}{937}$$

and

$$P[H \mid F^c] = \frac{P[H \cap F^c]}{P[F^c]} = \frac{108}{937} \div \frac{625}{937} = \frac{108}{625} = 0.173$$

This answers the question. ▲

2.12 Microeconomics, Macroeconomics, Finance, and Theory of Interest

This course dealt with four related fields of knowledge: microeconomics, macroeconomics, finance, and the theory of interest. Much if not all of this material is now integrated into the SOA courses FM (Financial Mathematics) (CAS Course 2) and MFE (Models for Financial Economics) (CAS Course 3F) with greater emphasis on applications.

Microeconomics

Microeconomics focuses on the role of individual firms and groups of firms with national and international economies. Key ideas of microeconomics are the demand and supply for individual goods and services, their trading and patterns of pricing, market equilibrium, and ideas such as the concepts of a monopoly, where one firm dominates the market, and an oligopoly, where a small number of firms dominate a national or global market. According to the SOA syllabus, actuaries should *be able to use microeconomic principles to build models to increase their understanding of the framework of contingent events and to use as a frame for activities such as pricing,* and *be able to use knowledge of microeconomic principles to increase their understanding of the markets in which we operate and of the regulatory issues.*

Macroeconomics

Macroeconomics deals with aggregate economic factors such as total national income and output, employment, balance of payments, rates of inflation, and the business cycle. One of the key ideas of macroeconomics is that of a gross national product: the total value of goods and services produced in an economy during a specified period time. According to the SOA syllabus, actuaries should understand macroeconomic principles to be able to develop economic models and assess the consequences of macroeconomics assumptions. They should understand *the relationship among interest rates, demand for money, consumption and investment using concepts such as the IS/LM (IS: investment = savings, LM: demand for money = supply of money) curve, fiscal and monetary policy, and how foreign exchange rates affect the gross national product and national income.* They should understand macroeconomic principles and know how to relate them to the business cycle.

Theory of Interest

The theory of interest is at the heart of actuarial science. It deals with the cost of money over time. According to the SOA syllabus, actuaries should understand how the theory of interest is used in annuity functions and be able to *apply the concepts of present and accumulated value for various streams of cash flows as a basis for future use in: reserving, valuation, pricing, duration, asset/liability management, investment income, capital budgeting, and contingencies.* Calculus plays a major role in the theory of interest since exponential functions, infinite series, and the continuous measurement of interest are key elements of financial modeling. Actuaries also need to be able to determine *the yield rates on investments and the time required to accumulate a given amount or repay a given loan amount,* and use annuity functions in financial context such as mortgages and similar products.

The starting point of actuarial science is that of an *annuity*, the idea of investing money, earning interest on the investment, and receiving payments in return. If you invest $1,000 at 5% interest per year, the bank will pay you an annuity of $50 per year to you or your heirs or until you withdraw your initial investment.

A fundamental variation on this theme is the idea of a *life annuity*. If you pay $1,000 to a life insurance company, the company may contract to pay you a fixed amount until you die. At that point the payments will cease and the initial investment is not refunded. The amount the company agrees to pay depends both on prevailing rates of interest and on how long the company expects you to live. Retirement benefits from pension plans are typical life annuities. This is the point where probability and statistics enter the picture. Most countries collect statistics on life expectancy and update this information on a period basis. The results are *life tables*. The so-called *Breslau Table* seems to be the first published record of this kind. In 1693, Edmond Halley published his analysis of the records of death of the city of Breslau in Germany (now Wroclaw, Poland). He started with a population of 1000 children of age 1, and calculated the number of survivors at different ages, up to the age of 84. Based on his table, Halley developed a method for calculating the premiums of life annuities dependent on two lives. One of the difficulties of using Halley's table for the purpose of calculating life annuities premiums was that the numbers in the table do not give rise to an obvious formula for mortality. A few years after Halley published his table, de Moivre tried to remedy this situation by postulating that the number of survivors decreased in arithmetical progression. If N is the initial population in any given year and d is the number of deaths per year, then the number of survivors k years later is $N - kd$. It turned out that de Moivre's assumption produced results that were sufficiently close to those of Halley to be of practical value.

Example Functions and Probabilities

Here are some examples of basic *annuity functions* essential for actuarial work.

Example 1 *Accumulated value of an annuity*

The function

$$S = Rs_{\overline{n}|i} = R\frac{(1+i)^n - 1}{i}$$

calculates the accumulated value S of an ordinary simple annuity of n payments of R dollars per payment. The expression

$$s_{\overline{n}|i} = \frac{(1+i)^n - 1}{i}$$

is known as the "accumulation factor for n payments," and is read as "s angle n at i." ▲

Example 2 *Discounted value of an annuity*

The function

$$A = Ra_{\overline{n}|i} = R\frac{1-(1+i)^{-n}}{i}$$

calculates the present value of the set of payments R due one period before the first payment. The expression

$$a_{\overline{n}|i} = \frac{1-(1+i)^{-n}}{i}$$

is known as the present value of an annuity-immediate of a payment of one for n periods, i.e., when the payment is made at the *end* of each period. ▲

Example 3 *Present value of an annuity due*

The expression

$$\ddot{a}_{\overline{n}|i} = (1+i)\,a_{\overline{n}|i}$$

is known as the present value of an *annuity due* of a payment of one for n periods, i.e., when the payment is made at the *beginning* of each period. ▲

Example 4 *Accumulated value of an annuity due*

The expression

$$\ddot{s}_{\overline{n}|i} = (1+i)\,s_{\overline{n}|i}$$

denotes the accumulated value of an *annuity due* of one at time n. ▲

Here are some typical probabilities and life functions used by actuaries.

Example 5 *Probability of death*

The expression

$$q_x$$

denotes the probability that an individual, alive at age x, will die before age $x+1$. ▲

Example 6 *Probability of survival*

The expression

$$p_x = 1 - q_x$$

denotes the probability that an individual, alive at age x, will survive beyond age $x+1$. ▲

Example 7 *Bounded probability of death*

The expression

$$_n q_x$$

denotes the probability that an individual, alive at age x, will die before age $x+n$. ▲

Example 8 *Bounded probability of survival*

The expression

$$_n p_x = 1 - {_n q_x}$$

denotes the probability that an individual, alive at age x, will survive beyond age $x+n$. ▲

Death and survival probabilities are used to define basic life insurance products. Here are some examples:

Example 9 *Pure endowment*

The function

$$_n E_x = (1+i)^{-n} {_n p_x}$$

computes the cost of an n-year endowment of one dollar to be paid to a person aged x years if that person reaches age $x+n$. ▲

Example 10 *Discounted value*

The function

$$a_x = \sum_{t=1}^{\infty} (1+i)^{-t} {_t p_x}$$

computes the discounted value of a one-dollar ordinary life annuity issued to someone of age x. ▲

Example 11 *Life annuity value*

The function

$$\ddot{a}_x = 1 + \sum_{t=1}^{\infty} (1+i)^{-t} {_t p_x}$$

computes the value of a one-dollar life annuity issued to someone of age x whose first premium payment is due now. ▲

Example 12 *Discounted value of a life annuity due*

The function

$$a_{x:\overline{n}} = \sum_{t=1}^{n} (1+i)^{-t} {_t p_x}$$

computes the discounted value of a one-dollar temporary life annuity due when n is the length of the payment period. ▲

Example 13 *Value of a life annuity due*

The function

$$\ddot{a}_{x:\overline{n}|} = \sum_{t=0}^{n-1} (1+i)^{-t} \, {}_t p_x$$

computes the value of a one-dollar life annuity due issued to someone of age x whose first premium payment is due now. ▲

Example 14 *Net single premium of term insurance*

The function

$$A^1_{x:\overline{n}|} = \sum_{t=0}^{n-1} (1+i)^{-(t+1)} \, {}_t p_x q_{x+t}$$

computes the net single premium for a one-dollar, n-year term insurance policy sold to a person x years old. ▲

Example 15 *Net single premium of whole life insurance*

The function

$$A_x = \sum_{t=0}^{\infty} (1+i)^{-(t+1)} \, {}_t p_x q_{x+t}$$

computes the net single premium for a one-dollar whole life insurance policy sold to a person x years old. ▲

Example 16 *Net single premium of an endowment*

The function

$$A^1_{x:\overline{n}|} + {}_n E_x$$

computes the net single premium for one-dollar, n-year endowment insurance policy. ▲

Example 17 *Annual premium of a whole life insurance policy*

The function

$$P_x = \frac{A_x}{\ddot{a}_x}$$

computes the net annual premium for a one-dollar whole life insurance policy. ▲

The Finance part of Course 2 [2002] deals with financial statements including balance sheets, income statements, and statements of cash flow. The main ideas involved are discounted cash flow, internal rate of return, present and future values of bonds, and applying the dividend growth model and price/earnings ratios concept to valuing stocks. Actuaries must be able to assess financial performance using net present value, the payback and discounted payback models, and internal rate of return and profitability index models. Among the key ideas are risk and return, and efficient markets. Actuaries must be able to valuate securities, apply measures of portfolio risk, and analyze the effects of diversification and systematic and unsystematic risks. They must be able to calculate portfolio risks, analyze the impact of individual securities on portfolio risks, identify efficient portfolios, and apply the CAPM [capital asset pricing model] to measure the cost of capital. They must also understand "the impact of financial leverage and long- and short-term financing policies on capital structure, sources of capital and the definitions of techniques for valuing basic options such as calls and puts."

Examination Topics

The sample examination consisted of 50 multiple-choice questions. They dealt with the following topics from the SOA and CAS syllabus:

Q1 Efficient market hypothesis, market prices, past price data, actively managed portfolios, semi-strong version of the efficient market theory, securities.

Q2 Free rider problem, public goods, private markets, governments, nonpaying consumers, social costs, positive prices, long-run marginal costs, long-run average costs.

Q3 Economies, contractionary phases, business cycles, indicators, downturn, unemployment rates, building permits, stock prices, delivery lags, business inventories, inventory accumulation.

Q4 Loans, amortizations, annual payments, effective rates, sinking funds.

Q5 Perpetuity-immediate annuities, present value.

Q6 Constant-cost competitive industries, long-run equilibrium, licensing fees, long-run market supply, long-run firm supply, fixed costs, prices, demand, average costs, marginal costs, output.

Q7 Loan, nominal interest rates, compound interest, lump sums, interest.

Q8 Shares, common stock, capital, earnings, treasury stock.

Q9 Consumer goods, marketplace, demand, prices, Engel curve, demand curve, personal income, income effect, substitution effect, compensated demand curve, normal goods, uncompensated demand curve, income elasticity, slopes, compensated price decline.

Q10 Investment, return level cash flow, internal rates of return, single cash flow, risk-free rates, market risk premiums, estimated beta, payback periods, net present value, annual cash flow, annuities, discount rates.

Q11 Salvage value, depreciation, declining balance method, sum-of-the-years digits method.

Q12 Investments, annual effective discount rates, interest.

Q13 Loans, present value, interest, principal.

Q14 Productivity, productivity growth, government spending, infrastructure, capital stock, real wage growth, workforce, demographic composition, labor, capital, economy, service jobs, real wages, real wage growth.

Q15 Production costs, delivery costs, equilibrium price, largest daily rate, outsourcing, profits.

Q16 Money supply, central bank, commercial banking system, public demand for currency, market interest rate, exogenous increase in interest rates, discount rate, reserve requirement ratio, bond sale, reserves.

Q17 Effective rates of interest, present value, perpetuity, present value, perpetuity-immediate.

Q18 Natural monopolies, prices, marginal costs, loss, fixed costs, marginal cost curves, industry demand curves, marginal revenue, average cost curves, competitive prices.

Q19 Net cash flow, opportunity costs, capital, net present value, expected economic income, cash flow.

Q20 Long-run real output, velocity of money, growth, monetary authority, target rates of inflation, money supply, price levels, growth rates.

Q21 Demand, supply, elasticity, exogenous increase in wages, prices, quantities, marginal production costs, equilibrium prices.

Q22 Investment, cash flow, after-tax weighted average costs of capital, net present value, equity financing, debt financing, marginal tax rates, equity costs, debt costs.

Q23 Real income, interest rates, economies, IS/LM framework, IS curves, LM curves, expansionary monetary policy, exogenous increase in domestic price levels, exogenous increase in savings, retirement, government spending, personal income tax, deficits.

Q24 Company assets, depreciation basis, marginal tax rates, after-tax rates, present value of tax shields.

Q25 Competitive firm, short-run operation, total revenue, total costs, fixed costs, average costs, marginal costs, average variable costs.

Q26 Investment, annual effective interest rates, accumulated values.

Q27 Stock prices, marginal requirements, marginal debts, interest, annual effective rates, dividends, return, short sale.

Q28 Monopolies, demand, marginal costs, prices, quantities, demand curves, continuous quantities.

Q29 Monopolies, marginal propensity to consume, income tax rates, government expenditure multiplier, exponential functions.

Q30 Stock prices, one-period put, exercise prices, risk-free rates, unexercised prices.

Q31 Investment, time-weighted returns, dollar-weighted returns.

Q32 Short-run supply curves, competitive industries, prices, industry output, production increase, industry supply curve, elastic supply curve, marginal cost curve, factor-price effect, shift down of the marginal cost curve.

Q33 Current liabilities, long-term liabilities, shareholder equity, total assets, EBIT [earnings before income and taxes], depreciation, interest, taxes, payout ratio, retained earnings, net income, dividends, internal growth rates.

Q34 Economies, goods, competitive supply and demand functions, prices, quantities, price ceilings, supply curve, deadweight loss, competitive equilibrium, consumer surplus, producer surplus, total surplus, excess demand.

Q35 Variance, equity returns, equal-weighted portfolio, beta, returns on assets, return on a market portfolio, slope, capital asset pricing model, derivatives (calculus).

Q36 Nominal exchange rates, inflation rates, real exchange rates.

Q37 Effective rates of interest, principals.

Q38 All-equity financed insurers, book value, return on equity, cash flow, annual earnings, dividends, free cash flow, opportunity costs, capital, discounted cash flow, plowback.

Q39 Debt ratio, debt beta, equity beta, expected return, risk-free interest rates, return on investment, target capital structure, risk, Modigliani-Miller capital structure theory, asset beta, capital asset pricing models.

Q40 Call option, common stocks, shares, standard deviations, continuous interest, compound interest, maturity of a call, risk-free rates, Black-Scholes, present value.

Q41 Bonds, semi-annual coupons, nominal yields, compound interest, annual effective interest rates, coupon payments, redemption value of bonds, annual effective yields, investment.

Q42 Supply and demand functions, prices, quantities, price elasticity of demand, initial equilibrium, percentage change, derivatives (calculus).

Q43 Market value, liabilities, debts, equity, beta, expected return, weighted average cost of capital, risk-free rates, expected risk premiums.

Q44 Earnings before interest and taxes, debt, corporate tax rate, dividend, average equity, return on average equity.

Q45 Force of interest, nominal rates of discount, convertible rates, accumulated value of funds, exponential functions, integrals.

Q46 Utility-maximizing consumers, indifference curves, utilities, slopes, budget curves.

Q47 Macro-economies, long-run view, real output, growth of real output, growth of inputs, velocity of money, growth in wage rates, wage-price spiral, inflation.

Q48 Stock prices, dividends, long-run dividend growth rates, capitalization rates, expected rates of return.

Q49 Investment, interest, nominal interest rates, convertible interest rates, simple interest, forces of interest, logarithmic functions.

Q50 Present value, annuities, perpetuity-immediate annuities, effective interest rates, annuity-immediate.

If we examine the frequency of some of the topics and techniques tested in this examination, we come up with the following result: price (16/50), marginal (13/50), cost (10/50), interest (10/50), growth (9/50), present value (9/50), cash flow (8/50), curve (8/50), effective (8/50), investment (7/50), market (7/50), debt (6/50), return (6/50), stock price (6/50), demand (5/50), dividend (5/50), equity (5/50), expected value (5/50).

Examination Questions and Answers

Here are some examples from a sample examination that show how some of these ideas and techniques were tested. The cited questions involve a variety of ideas, ranging from supply and demand, the business cycle, money supply, marginal tax rates, to effective interest rates, stock prices, and the valuation of companies. The questions also use two special actuarial symbols: the symbol $a_{\overline{n}|}$, which stands for the value of an annuity of one dollar per year for n years, payable at the end of each year, and the symbol $a_{\overline{n}|i}$, which denotes the value of an annuity of one dollar per year for n years at i percent interest per year, payable at the end of each year.

Question 1 *Suppose the economy is entering the contractionary phase of a business cycle. Which of the following is an indicator of this downturn in economic activity? (1) A decrease in the unemployment rate. (2) An increase in the number of new building permits for private housing units. (3) An increase in stock prices. (4) An increase in delivery lags. (5) An increase in business inventories.*

Answer An increase in business inventories indicates that demand is not as high as businesses anticipated, resulting in inventory accumulation. The decrease in demand is a reflection of the downturn in economic activity. ▲

Question 2 *A 20-year loan of 20,000 may be repaid under the following two methods: (1) Amortization method with equal annual payments at an annual effective rate of 6.5%, (2) Sinking fund method in which the lender receives an annual effective rate of 8% and the sinking fund earns an annual effective rate of j. Both methods require a payment of X to be made at the end of each year for 20 years. Calculate j.*

Answer We note that

$$X = \frac{20000}{a_{\overline{20}|0.065}} = 1815.13$$

Therefore,

$$1815.13 = \frac{20000}{s_{\overline{20}|j}} + (0.08)(20,000)$$

$$s_{\overline{20}|j} = 92.97$$

$$j = 14.18\%$$

This answers the question. ▲

Question 3 *Suppose a constant-cost, competitive industry is in long-run equilibrium. Now suppose the government imposes an annual licensing fee as a requirement for firms to produce in the industry. As a result of this fee, what will happen to the quantity supplied in the market and the quantity supplied by an individual firm in the long run?*

The possible answers are

(A) The quantity supplied in the market will increase, and the quantity supplied by an individual firm will increase.

(B) The quantity supplied in the market will increase, and the quantity supplied by an individual firm will decrease.

(C) The quantity supplied in the market will decrease, but the quantity supplied by an individual firm will not change because some firms go out of business.

*(D) The quantity supplied in the market will decrease, and the quantity supplied
by an individual firm will decrease.*

*(E) The quantity supplied in the market will decrease, and the quantity supplied
by an individual firm will increase.*

Answer The licensing fee works the same as an increase in fixed costs; it shifts
the market supply upward, increasing price and decreasing quantity demanded.
At the firm level, however, it increases average costs without changing marginal
costs; therefore, the representative firm increases output. This apparent paradox is
resolved by the fact that in the long run some firms will go out of business. ▲

Question 4 *Bruce and Robbie each open up new bank accounts at time 0. Bruce
deposits 100 into his bank account, and Robbie deposits 50 into his. Each account
earns an annual effective discount rate of d. The amount of interest earned in
Bruce's account during the 11th year is equal to X. The amount of interest earned
in Robbie's account during the 17th year is also equal to X. Calculate X.*

Answer Bruce's interest in the 11th year is

$$\frac{100}{(1-d)^{10}}\left[\frac{1}{(1-d)}-1\right]=X$$

and Robbie's interest in the 17th year is

$$\frac{50}{(1-d)^{16}}\left[\frac{1}{(1-d)}-1\right]=X$$

$$=\frac{100}{(1-d)^{10}}\left[\frac{1}{(1-d)}-1\right]$$

$$(1-d)^6=\frac{1}{2}\implies d=10.91\%$$

$$X=\frac{100}{(1-0.1091)^{10}}\left[\frac{1}{1-0.1091}-1\right]=38.88$$

This answers the question. ▲

Question 5 *The money supply is determined by the combined actions of the central bank, the commercial banking system, and the public's preferences regarding
how they hold money. Which of the following will result in an increase in the
money supply? (1) An increase in the public's demand for currency. (2) An exogenous increase in market interest rates. (3) The central bank increases the discount
rate. (4) The central bank increases the reserve requirement ratio. (5) The central
bank sells bonds to the public.*

Answer An increase in market interest rates will result in banks lending out excess
reserves, which lowers free reserves and increases the money supply. ▲

Question 6 *At an annual effective interest rate of i, i > 0%, the present value of a perpetuity paying* 10 *at the end of each 3-year period, with the first payment at the end of year* 6, *is* 32. *At the same annual effective rate of i, the present value of a perpetuity-immediate paying* 1 *at the end of each 4-month period is X. Calculate X.*

Answer We note that

$$v^3 \frac{10}{(1+i)^3 - 1} = 32$$

Therefore,

$$10v^3 = 32(1+i)^3 - 32$$

Multiplying both sides by $(1+i)^3$ yields

$$10v^3(1+i)^3 = 32(1+i)^6 - 32(1+i)^3$$

and since $v^3(1+i)^3 = 1$, we have

$$0 = 32(1+i)^6 - 32(1+i)^3 - 10$$

This tells us that

$$(1+i)^3 = \frac{32 \pm \sqrt{2304}}{64} = 1.25$$

$$i = 7.72\%$$

$$X = \frac{1}{(1+i)^{1/3} - 1}$$

$$= \frac{1}{(1.0772)^{1/3} - 1} = 39.84$$

This answers the question. ▲

Question 7 *A company invests* 20,000 *in a project. The project is expected to have cash flows of* 3000 *at the end of each year for* 15 *years, with the first cash flow expected one year after the initial investment. Using the project's after-tax weighted average cost of capital, the project has a net present value of* 2496.27. *The following gives additional information about the company: (1) The company is financed with 40% equity and 60% debt. (2) The company's marginal tax rate is 25%. (3)* $r_E = 2r_D$, *where* r_E *is the cost of equity and* r_D *is the cost of debt. Calculate* r_E.

Answer Let i denote the after-tax weighted average of capital. Then

$$3000a_{\overline{15}|i} - 20,000 = 2496.27$$

Therefore, $a_{\overline{15}|i} = 7.49876$. Hence $i = 10.25\%$. It follows that

$$10.25 = r_E\,(0.4) + r_D\,(1 - T_x)\,(0.6)$$

$$= 0.4 \cdot r_E + \frac{1}{2}r_E\,(1 - 0.25)\,(0.6) = 0.4r_E + 0.225r_E.$$

Hence $r_E = 16.4\%$. ▲

Question 8 *A company has an asset with a depreciation basis of* 100,000 *which can be depreciated by the following schedule:*

Year	Percent
1	33.33
2	44.45
3	14.81
4	7.41

The marginal tax rate is 35% and the pretax borrowing rate is 12%. Calculate the present value of the tax shields created by the depreciation.

Answer From the given information we conclude that

	Year 1	Year 2	Year 3	Year 4
Dollar deductions	33,330	33,340	14,810	7,410
Tax shields	11,666	15,558	5,184	2,594

and the after tax rate is $0.12(0.65) = 0.078$. Hence

$$PV = 11{,}666/1.078 + 15{,}558/1.078^2 + 5{,}184/1.078^3 + 2{,}594/1.078^4$$
$$= 30{,}267$$

This answers the question. ▲

Question 9 *A stock price can go up by 20% or down by 15% over the next period. The current stock price is greater than 70. You own a one-period put on the stock. The put has an exercise price of 78.26. The risk-free rate is 11.25%. If the put is exercised today, the amount received will be X. The price of the put today (unexercised) is also X. Calculate the current stock price.*

Answer Let p be the probability that a stock price goes up. Then

$$20p + (-15)\,(1 - p) = 11.25 \rightarrow p = 0.75$$

Moreover, the put value is

$$78.26 - X$$

if exercised now, and

$$\frac{0.75 \cdot 0 + 0.25 \cdot (78.26 - 0.85X)}{1.1125}$$

if not exercised now. By equating these two expressions and solving for X, we get $X = 75$. ▲

Question 10 *Which of the following statements about the short-run supply curve for a competitive industry is false? (1) As price rises, industry output goes up because firms in the industry increase production. (2) As price rises, firms not previously producing will start up production and thereby further increase industry output. (3) As price rises, entry of new firms tends to make the industry supply curve more elastic than the supply curve of typical firms in the industry. (4) As price and output increase for the industry, the factor-price effect is likely to make the industry supply curve less elastic. (5) As price and output increase for the industry, the marginal cost curve of each firm in the industry will likely shift down because of the factor-price effect.*

Answer As a result of the factor-price effect, the marginal cost curves of the firms do not shift down but up. ▲

Question 11 *You are given:*

Current Liabilities	300	*EBIT*	400
Long-Term Liabilities	700	*Depreciation*	100
Shareholder Equity	1400	*Interest*	50
Total Assets	2400	*Taxes*	60

The company's payout ratio is 10%. Determine the company's internal growth rate.

Answer The following calculations show that the internal growth rate is 10.875%:

1. Assets = Liabilities + Shareholder Equity:

$$300 + 700 + 1400 = 2400$$

2. Net Income = EBIT − Interest − Taxes:

$$400 - 50 - 60 = 290$$

where the depreciation has already been subtracted to get EBIT.

3. Payout Ratio:

$$0.10 = \frac{\text{Dividends}}{\text{Net Income}}$$
$$= \frac{\text{Dividends}}{290}$$

Therefore,

$$\text{Dividends} = 29$$

Putting it all together, we have

$$\text{Retained Earnings} = \text{Net Income} - \text{Dividends}$$
$$= 290 - 29 - 261$$

and

$$\text{Internal Growth Rate} = \frac{\text{Retained Earnings}}{\text{Assets}}$$
$$= \frac{261}{2400} = 10.875\%$$

This answers the question. ▲

Question 12 *You are the chief actuary for a small, all-equity financed insurer. The current book value of equity is* 1000. *In years 1 and 2, you will earn a return on equity [ROE] of 20% and reinvest all earnings. Starting in year 3 (and every year thereafter), your company's ROE will be 15%, your free cash flow will be 50% of annual earnings, and you will pay a dividend equal to 100% of free cash flow. You have been approached by another insurer who would like to buy your company. Assuming an opportunity cost of capital equal to 15%, use discounted cash flow to find the value of your company.*

	Y1	Y2	Y3	Y4	Y5
Book Equity	1000	1200	1440	1548	1664.1
ROE	20%	20%	15%	15%	15%
Earnings	200	240	216	232.2	249.62
Dividends	0	0	108	116.1	124.81
Plowback	200	240	108	116.1	124.81
Free Cash Flow	0	0	108	116.1	124.81

Answer Starting in Year 3, we have

$$\text{Dividend Growth Rate} = \text{Plowback} \cdot \text{ROE}$$
$$= (0.5)(0.15) = 0.075$$

Therefore,

$$PV @ t = 2 \text{ of Future Dividends} = PV @ t = 2 \text{ of Free Cash Flow}$$
$$= 108/(0.15 - 0.075) = 1440$$

and

$$PV @ t = 0 \text{ of Free Cash Flow}$$
$$= 1440 (1.15)^{-2} \approx 1089$$

This answers the question. ▲

Question 13 *You are interested in purchasing a call option on a common stock that is currently trading at a price of* 100 *per share. You are given the following information: (1) The standard deviation of the continuously compounded annual rate of return on the stock is* 0.4. *(2) The time to maturity of the call is three months. (3) At the risk-free rate,*

$$\ln \left(\frac{Current\ Share\ Price}{Present\ Value\ of\ the\ Exercise\ Price} \right) = -0.08$$

Calculate the price of each call option using Black-Scholes.

Answer The Black-Scholes model tells us that the price of a call option is the difference between the expected benefit from acquiring the stock outright and the present value of paying the exercise price on the expiration days. In other words

$$Price = CP \times N(d_1) - PV \times N(d_2)$$

where the current price CP is 100, and where the present value PV of the exercise price is 108.33, since at the risk-free rate and a current share price of 100, the log of the ratio of the current share price and the present value of the exercise price is 0.08. Moreover, $d_1 = -0.3$ and $d_2 = 0.5$, since $t = 0.25$ and $\sigma = 0.4$, so that $\sigma \times \sqrt{(t)} = 0.2$. Therefore,

$$Price = 100 \times N(-0.3) - 108.33 \times N(-0.5)$$
$$= 100 \times 0.3821 \quad 108.33 \times 0.3085$$
$$= 4.79$$

This answers the question. ▲

Question 14 *The supply and demand functions for a good are* $P = 1 + 4Q$ *and* $P = 4 - 2Q$, *respectively, where P is price and Q is quantity. Now suppose an increase in the price of an input causes the supply function to become*

$$P = 2 + 4Q$$

What is the price elasticity of demand at the initial equilibrium?

Answer The correct answer follows from the definition of the price elasticity of demand. The percentage change in price from the initial equilibrium is $1/9$, and the percentage change in quantity demanded is $-1/3$; hence the price elasticity of demand is -3.00. ▲

Question 15 *A company's stock is currently selling for 28.50. Its next dividend, payable one year from now, is expected to be 0.50 per share. Analysts forecast a long-run dividend growth rate of 7.5% for the company. Tomorrow the long-run dividend growth rate estimate changes to 7%. Calculate the new stock price.*

Answer The current capitalization rate is

$$P_0 = DIV_1/(r-g)$$

In other words,

$$28.50 = 0.50/(r-0.075)$$

Therefore, $r - 0.075 = 0.50/28.50 = 0.0175$, so that $r = 0.0925$. When the long-run growth rate changes, current price should adjust to reflect this change, and to keep the expected rate of return constant. This tells us that

$$P_0 = 0.50/(0.0925 - 0.07) = 22.22$$

This answers the question. ▲

2.13 | Use of Actuarial Models

This course dealt with the use of actuarial models. In the Basic Education Catalog, Spring 2003 (see [4]), the Society of Actuaries describes the learning objective of the course by saying that *this course develops the candidate's knowledge of the theoretical basis of actuarial models and the application of those models to insurance and other financial risks.* The word "model" is used by scientists for the tools they have developed to describe and explore their environments. Physicists build models to understand the universe around us, biologists build models to understand the long-term dynamics of interacting populations, economists build models to understand the interaction of the supply and demand of consumer goods, and actuaries build models to analyze the profitability of insurance plans, pension plans, and the returns on investment portfolios. Most scientists use their models to predict some aspects of the future and to provide a basis on which decisions can be made. These decisions can be ecological, economic, commercial, and financial. In the case of actuaries, the decisions are usually financial.

In collaboration with Wolfram Research, the makers of the Mathematica software package, and ACTEX, the providers of actuarial study tools, Bruce Jones of the University of Western Ontario has developed a beautiful interactive course for studying actuarial models and building them (see [12]). This course is ideal for preparing for the examinations in Courses 3 [2002] and 4 [2002]. Jones begins his exposition by pointing out that "from the perspective of the actuary, a model can be defined as *a mathematical representation of a phenomenon.* This phenomenon usually has financial implications. Examples of phenomena that actuaries frequently model include the following: the time until death of an individual insured under a life insurance policy, the amount of insured losses under a health, automobile or property insurance policy, and the return of an investment portfolio."

From a mathematical point of view, a model can be many things. Among the familiar models are graphs that help us visualize quantitative relationships and functions that capture changes in a phenomenon and allow us to make predictions about the future.

The starting points for building actuarial models are historical data and probabilistic assumptions. For example, a frequently encountered mathematical model in actuarial science is the Poisson probability distribution. It is used, for example, to model phenomena such as the number of automobile accidents at a particular intersection in a city over a fixed period of time.

The basic idea from statistics needed here is that of a *random variable*. What are random variables? In an attempt to simplify their definition, many authors have different ways of defining them. One author writes that "a random variable is a variable whose values are determined by chance." Another writes that "a random variable is a real-valued function for which the domain is a sample space." At some point, all authors distinguish between discrete and continuous random variables. The common element here is that a random variable is above all a *real-valued function*. Moreover, its domain has a certain structure that statisticians refer to as a *sample space*. The elements of the sample space are known as *sample points* and sets of sample points are called *events*. In Ott and Longnecker's book on statistical methods (see [17]), the random variables of interest in Course 3 [2002] are called *quantitative random variables*. In the discrete case, they allow us to introduce quantitative measures such as means and standard deviations over their range of values. The key actuarial idea associated with a random variable is that of the *expected value* of the variable.

Example 1 *Discrete random variable*

If you take two coins and list the number of possible head-tail combinations that can be obtained by tossing the coins, the result is a sample space S. The events are

$$e_1 = HH, e_2 = HT, e_3 = TH, e_4 = TT$$

The function $Y : S \to \mathbb{R}$ defined by

$$Y(e_1) = 0, Y(e_2) = Y(e_3) = 1, Y(e_4) = 2$$

is a random variable. It counts the number of heads of each sample point. Since the domain of Y is finite, the function is called *discrete*. ▲

The next required idea is that of a *probability distribution* of a random variable. It assigns to each value x of the random variable X a probability $0 \le p(x) \le 1$, also denoted by $P(X = x)$, that measures the likelihood that the value x is attained. It is assumed that the sum of all values $p(x)$ over the domain of X is 1.

Example 2 *Continuous random variable*

The change in earnings per share of a particular stock over a fixed period of time is a random variable. The sample space is an interval on the real line marking off the time period over which the change is measured. The variable is continuous since it can take on arbitrary real numbers as values at all points of time in the interval. ▲

Example 3 *Expected value*

Suppose that you would like to insure your laptop computer for $2,000 against theft for one year. Suppose further that an insurance company has empirical evidence that the probability of having the laptop stolen in the first year is 1/10. What is your expected return from the insurance company if the premium you are charged is $100?

You have a chance of 1/10 of receiving $1,900 from the insurance company since you have already paid the company $100 in premiums. On the other hand, you have a chance of 9/10 of losing the $100 you have paid. The expected value of the probability distribution for X is

$$E(X) = (1900) \times (1/10) + (-100) \times (9/10) = 100$$

This means that if you insure your computer with the given company over a number of years, you will have an average net gain of $100 per year. The expected value from the insurance company's point of view, on the other hand, is

$$E(Y) = (-1900) \times (1/10) + (100) \times (9/10) = -100$$

In other words, the company can expect to lose $100 on average on this policy. ▲

Probability distributions are important statistical tools for analyzing the properties of random variables. In actuarial science, the binomial distributions and their associated Poisson and normal distributions play an important role.

Definition 4 *The function $P(y)$ is actually the limiting value of the widely known binomial probability distribution. It is derived from the binomial distribution by noting that*

$$\lim_{n \to \infty} \left(1 - \frac{\lambda}{n} \right)^n = e^{-\lambda}$$

and that therefore, with $\lambda = np$,

$$\lim_{n \to \infty} \binom{n}{y} p^y (1-p)^{n-y} = \frac{\lambda^y}{y!} e^{-\lambda}$$

Let us illustrate the descriptive and predictive aspect of a mathematical model by recalling a classical illustration of the binomial distribution due to Weldon. For historical details, we refer to the beautiful article on probability in the 1911 edition of the *Encylopedia Britannica* (see [10], page 394).

Example 5 *The Weldon experiment*

Suppose that we have n independent events, that the probability of a successful outcome of an event is p, and the probability of an unsuccessful event is q. If N represents the number of trials, then the formula

$$N(p+q)^n$$

counts the probable frequencies of the different results in a given number of trials.

Suppose that twelve dice are thrown a certain number of times, and that each face showing a 4, 5, or 6 is considered a success, whereas each face showing a 1, 2, or 3 is considered a failure. Then the probabilities of success and failure of each throw of the twelve dice is 1/2. Then

$$N\left(\frac{1}{2}+\frac{1}{2}\right)^{12}$$

where N is the total number of throws. Moreover, the binomial expansion of the expression

$$\left(\frac{1}{2}+\frac{1}{2}\right)^{12}$$

yields $A+B$, where

$$A = \frac{1}{4,096} + \frac{12}{4,096} + \frac{66}{4,096} + \frac{220}{4,096} + \frac{495}{4,096} + \frac{792}{4,096}$$

and

$$B = \frac{924}{4,096} + \frac{792}{4,096} + \frac{495}{4,096} + \frac{220}{4,096} + \frac{66}{4,096} + \frac{12}{4,096} + \frac{1}{4,096}$$

Let $N = 4,096$. Then

$$4,096\,(A+B)$$

is the sum of the theoretical frequencies of the different possible successes of 4,096 throws of twelve dice.

The following table compares these frequencies with the experimental frequencies found by Weldon:

Successes	Observed Frequencies	Theoretical Frequencies
0	0	1
1	7	12
2	60	66
3	198	220
4	430	495
5	731	792
6	948	924
7	847	792
8	536	495
9	257	220
10	71	66
11	11	12
12	0	1
Total	4,096	4,096

We can see that the relationship between the two distributions is very close. From an actuarial point of view, one of the important properties of the binomial distribution is the fact that it is a building block for the Poisson distribution.

Here are a number of typical examples illustrating the use of the binomial, the Poisson, and the normal distribution:

Example 6 *A binomial experiment*

Suppose that a student is writing a multiple-choice examination consisting of 40 questions, each with five possible choices. Calculate the probability that the student guesses exactly 20 right answers.

Answer The probability of success in a binomial experiment with x successes in n trials is given by the formula

$$P(x) = \frac{n!}{(n-x)!x!} p^x q^{(n-x)}$$

where p is the probability of success in a single trial, and q is the probability of failure in a single trial. Since $n = 40$, $x = 20$, and $p = \frac{1}{5}$, we have

$$P(20) = \frac{40!}{20!20!} \left(\frac{1}{5}\right)^{20} \left(\frac{4}{5}\right)^{20} \approx 1.66650^{-5}$$

This answers the question. ▲

Example 7 *A Poisson experiment*

In a clinical trial, 1,000 patients were treated with a new drug. Suppose that the known probability p of a person experiencing negative side effects is 0.0025. What is the probability that none of the 1,000 patients participating in the trial experience negative side effects?

Answer According to the Poisson formula, the probability of y successes in n trials is given by the formula

$$P(y) = \frac{\lambda^y}{y!}e^{-\lambda}$$

where $y = 0$ and $\lambda = np = 1000 \times 0.0025 = 2.5$. Therefore,

$$P(0) = \frac{2.5^0}{0!}e^{-2.5} \approx 0.082085$$

This answers the question. ▲

This example could of course have been answered more directly by noting that the question is asking for the probability of 1,000 failures in 1,000 trials of an experiment with success rate 0.0025. This is just

$$(1 - .0025)^{1000} \approx 0.81828$$

Example 8 *A normal approximation*

Whatever its beauty and theoretical correctness as a model for statistical analysis, the binomial distribution is often computationally too complex for practical use. Consider the following problem, discussed in detail in Ott and Longnecker's book on statistical methods (see [17], page 182). A thousand voters are polled to determine their opinion on a municipal merger. What is the probability that 460 or fewer of them favor the merger if it is assumed that 50% of the entire population favors the change?

Answer In the binomial experiment, $n = 1,000$, and the probability $p = 1/2$. To answer the questions, we must compute the sum

$$P = P(460) + P(459) + \cdots + P(1) + P(0)$$

where

$$P(460) = \frac{1000!}{460!540!}\left(\frac{1}{2}\right)^{460}\left(\frac{1}{2}\right)^{540}$$

and so on. The number of calculations required for the solution is enormous. What is our way out? The central limit theorem for sums (see [17], Appendix G, for example) enables us to approximate P using an approximate normal curve as an approximation to the required binomial distribution.

The graphs of the functions

$$f(y) = \frac{1}{\sqrt{2\pi}\sigma} e^{-\frac{(y-\mu)^2}{2\sigma^2}}$$

are bell-shaped curves known as *normal curves*, whose shape depends on the parameters μ and σ. Moreover, the area

$$P(a \leq Y \leq b) = \int\limits_{a}^{b} f(y) \, dy$$

under the curve of f can be interpreted as a probability.

If μ is the mean and σ the standard deviation of a normally distributed random variable Y with density function $f(y)$, then the probability that a randomly chosen value of Y will lie between a and b is $P(a \leq Y \leq b)$.

It is explained in Ott and Longnecker's book on statistical methods (see [17], page 184) that for large n and p not near 0 or 1, the distribution of a binomial random variable y may be approximated by a normal distribution with $\mu = np$ and $\sigma^2 = np(1-p)$, provided that $np \geq 5$ and $n(1-p) \geq 5$. The polling problem can be solved using the normal distribution since $np = 1000 \times 0.5 = 500 = n(1-p) \geq 5$.

Since most integrals involved in normal distribution problems have no closed-form solutions, approximate values of the integrals have been tabulated. For this purpose, an additional simplification has been introduced. Every normal distribution can be converted to standard form by letting

$$z = \frac{y - \mu}{\sigma}$$

and looking the value of z up in a table for standard normal curve areas.

If $\mu = np = 500$ and $\sigma = \sqrt{np(1-p)} = \sqrt{250} = 15.811$, then

$$z = \frac{y - \mu}{\sigma} = \frac{460 - 500}{15.811} \approx -2.53$$

Table 1 in the Appendix of Ott and Longnecker's book on statistical methods (see [17]) tells us that the area under the normal curve to the left of 460 (for $z = -2.53$) is 0.0057. Therefore, the probability of observing 460 or fewer favoring the merger is about 0.0057. ▲

Examination Topics

The sample examination consists of forty multiple-choice questions. It deals with the following topics from the SOA and CAS syllabus:

Q1 Survival function, de Moivre's law, limiting age, integrals.

Q2 Term insurance, benefits, premiums, loss random variable.

Q3 Random variables, gamma distributions, variances, means, Poisson distributions, expected values, negative binomial distributions.

Q4 Poisson distributions, random variables, variances, compound distributions, independent processes.

Q5 Life insurance, whole life policy, level annual benefit premiums, benefit reserves.

Q6 Multiple decrement models, life tables, exponential functions, probabilities, integrals.

Q7 Probability models, probabilities, expected value, Markov chain.

Q8 Stock prices, geometric Brownian motion, drift coefficients, mean, variance, inverse transform method, uniform distribution, random numbers, exponential functions.

Q9 Life insurance, fully discrete insurance, annual benefit premiums, life expectancy.

Q10 Multiple decremental models, expected values, exponential functions, logarithmic functions, integrals.

Q11 Term insurance, death benefits, inverse transform method, present value random variable, uniform distributions.

Q12 Life insurance, death and surrender benefits, mortality tables, surrender rates, inverse transform method, policy terminated by death, policy terminated by surrender, uniform distributions, random variable indicating death, random variable indicating lapse of policy.

Q13 Time-until-death, hyperbolic assumption at fractional ages, independent lives, probabilities.

Q14 Life-table functions, force of mortality, mortality graphs.

Q15 Automobile insurance, negative binomial distributions, means, variances, Poisson distributions, gamma distributed means, variance of gamma distributions.

Q16 Probability distributions, mean, variance, independence, normal approximations, expected values.

Q17 Term insurance, present value random variable, death benefits, actuarial present value.

Q18 Endowment insurance, discrete insurance, death benefits, maturity benefits, level annual benefit premiums, benefit reserves, actuarial present value, future benefits.

Q19 Stop-loss insurance, independence, loss distribution, deductibles, actuarial expected value.

Q20 Insurance claims, compound Poisson claims process, probability, moment-generating functions, continuous premium rates, adjustment coefficients, exponential functions, expected values.

Q21 Markov process, insurance claims, probabilities of claims, independence, dividends, probability of failure.

Q22 Markov process, insurance claims, probabilities of claims, independence, expected dividends.

Q23 Continuous two-life annuities, continuous single life annuities, actuarial present value.

Q24 Disability insurance, length of payment random variable, gamma distributions, actuarial present value, improper integrals, exponential functions.

Q25 Discrete probability distributions, recursion relations, Poisson distributions, exponential functions, factorial function.

Q26 P/C insurance, loss models, aggregate loss, compound Poisson distributions, expected value, exponential distributions, deductibles, memoryless property.

Q27 Mortality models, expected number of survivors, uniform distribution of deaths (UDD), constant force assumptions.

Q28 Time-until-death, force of mortality, uniform distributions, probability of death, expected value, improper integrals, exponential functions.

Q29 Loss models, probability distributions, expected value, standard deviation, variance.

Q30 Stop-loss insurance, security loads, probability distributions, deductibles, expected value, sums of independent random variables.

Q31 Term insurance, level benefit premiums, benefit reserves, actuarial present value.

Q32 Whole life insurance, fully continuous insurance, level premiums, equivalence principle, death benefits, interest rates, loss random variable, future lifetime random variable.

Q33 Mortality models, uniformly distributions, complete-expectation-of-life, integrals.

Q34 Whole life insurance, death benefits, premiums, mortality, life tables, minimum annual rates of return, investments.

Q35 Whole life insurance, actuarial present value, force of mortality, death benefits, future lifetimes, independence, common stock model.

Q36 Poisson distributions, mean, probability, variance, compound distributions, expected value.

Q37 Poisson process, intensity functions, independence, distributed random variables, uniformly distributed claims, number of claims as a random variable, conditional expected value, integrals.

Q38 Whole life insurance, probabilities, death benefits, level premiums, independence, mortality, life tables, equivalence principle, benefit reserves, actuarial present value, future benefits.

Q39 Annuities, mortality, life tables, independence, normal approximations, present value random variables, lives, standard deviations.

Q40 Whole life insurance, death benefits, benefit premiums, mortality, life tables.

If we examine the frequency of some of the topics and techniques tested in this examination, we come up with the following result: insurance (23/40), benefit (19/40), random variable (19/40), distribution (18/40), expected value (14/40), function (12/40), probability (12/40), independence (10/40), mortality (9/40), premium (9/40), life (8/40), mean and standard deviation (8/40), exponential function (7/40), integration (7/40), variance (7/40), actuarial present value (6/40). The mathematical and statistical ideas in the course include binomial distributions, Poisson distributions, normal distributions, the analysis of stock prices, and the calculation of insurance premiums. A detailed course description such as the one published by the Society of Actuaries in its Basic Education Catalog, Fall 2002 (see [3]), for example, completes the picture.

Examination Questions and Answers

Here are some examples from a sample examination showing how some of these ideas and techniques were tested.

Question 1 *For a given life age 30, it is estimated that an impact of a medical breakthrough will be an increase of 4 years in*

$$\overset{\circ}{e}_{30}$$

the complete expectation of life. Prior to the medical breakthrough, $s(x)$ followed de Moivre's law with $w = 100$ as the limiting age. Assuming de Moivre's law still applies after the medical breakthrough, calculate the new limiting age.

Answer By de Moivre's law,

$$\overset{\circ}{e}_{30} = \int_0^{w-30} \left(1 - \frac{t}{w-30}\right) dt = \left[t - \frac{t^2}{2(\omega-30)}\right]_0^{\omega-30} = \frac{w-30}{2}$$

Prior to the medical breakthrough, with $w = 100$, we therefore have

$$\overset{\circ}{e}_{30} = \frac{100-30}{2} = 35$$

After the medical breakthrough,

$$\left(\overset{\circ}{e}\right)_{30}' = \overset{\circ}{e}_{30} + 4 = 39 = \frac{w'-30}{2}$$

It follows that $w' = 108$. ▲

Question 2 *On January 1, 2002, Pat, age 40, purchases a 5-payment, 10-year term insurance of 100,000: (1) Death benefits are payable at the moment of death. (2) Contract premiums of 4000 are payable annually at the beginning of each year for 5 years. (3) $i = 0.05$. (4) L is the loss random variable at time of issue. Calculate the value of L if Pat dies on June 30, 2004.*

Answer It follows from the given information that

$$_0L = 100{,}000v^{2.5} - 4000\ddot{a}_{\overline{3}|.05} = 77{,}079$$

This answers the question. ▲

Question 3 *For a fully discrete 20-payment whole life insurance of 1000 on (x), you are given: (1) $i = 0.06$. (2) $q_{x+19} = 0.01254$. (3) The level annual benefit premium is 13.72. (4) The benefit reserve at the end of year 19 is 342.03. Calculate 1000 P_{x+20}, the level annual benefit premium for a fully discrete whole life insurance of 1000 on $(x+20)$.*

Answer The given information tells us that

$$1000 \, {}^{20}_{20}V_x = 1000A_{x+20}$$

$$= \frac{1000 \left({}^{20}_{19}V_x + {}_{20}P_x \right)(1.06) - q_{x+19}(1000)}{p_{x+19}}$$

$$= \frac{(342.03 + 13.72)(1.06) - 0.01254(1000)}{0.98746}$$

$$= 369.18$$

and, therefore,

$$\ddot{a}_{x+20} = \frac{1 - 0.36918}{0.06/1.06} = 11.1445$$

It follows that

$$1000P_{x+20} = 1000\frac{A_{x+20}}{\ddot{a}_{x+20}} = \frac{369.18}{11.1445} = 33.1$$

This answers the question. ▲

Question 4 *A coach can give two types of training, "light" or "heavy," to his sports team before a game. If the team wins the prior game, the next training is equally likely to be light or heavy. But, if the team loses the prior game, the next training is always heavy. The probability that the team will win the game is 0.4 after light training and 0.8 after heavy training. Calculate the long run proportion of time that the coach will give heavy training to the team.*

Answer Let "light training" be State 1 and "heavy training" be State 2. Then the probabilities P_{ij} involved are

$$P_{11} = 0.4 \times 0.5 + 0.6 \times 0 = 0.2$$
$$P_{12} = 0.4 \times 0.5 + 0.6 = 0.8$$
$$P_{21} = 0.8 \times 0.5 + 0.2 \times 0 = 0.4$$
$$P_{22} = 0.8 \times 0.5 + 0.2 = 0.6$$

and the transition matrix of the given Markov process is therefore

$$P = \begin{bmatrix} 0.2 & 0.8 \\ 0.4 & 0.6 \end{bmatrix}$$

Let π_1 be the long-run probability that light training will be given, and π_2 that heavy training will take place. Then we can tell from the matrix P that

$$\pi_1 = 0.2\pi_1 + 0.4\pi_2$$
$$\pi_2 = 0.8\pi_1 + 0.6\pi_2$$

We also know that $\pi_1 + \pi_2 = 1$. Hence

$$1 - \pi_2 = 0.2 (1 - \pi_2) + 0.4\pi_2$$
$$= 0.2 + 0.2\pi_2$$

Therefore, $1.2\pi_2 = 0.8$ and $\pi_2 = \frac{8}{12} = \frac{2}{3}$. ▲

Question 5 (x) *and* (y) *are two lives with identical expected mortality. You are given that* $P_x = P_y = 0.1$, *that* $P_{\overline{xy}} = 0.06$, *where* $P_{\overline{xy}}$ *is the annual benefit premium for a fully discrete insurance of 1 on* $(\overline{xy})$, *and that* $d = 0.06$. *Calculate the premium* P_{xy}, *the annual benefit premium for a fully discrete insurance of 1 on* (xy).

Answer We note that $P_s = 1/\ddot{a}_s - d$, where s can stand for any of the statuses under consideration.

Therefore,

$$\ddot{a}_s = \frac{1}{P_s + d}$$

$$\ddot{a}_x = \ddot{a}_y = \frac{1}{0.1 + 0.06} = 6.25$$

$$\ddot{a}_{\overline{xy}} = \frac{1}{0.06 + 0.06} = 8.333$$

and since $\ddot{a}_{\overline{xy}} + \ddot{a}_{xy} = \ddot{a}_x + \ddot{a}_y$,

$$\ddot{a}_{xy} = 6.25 + 6.25 - 8.333 = 4.167$$

$$P_{xy} = \frac{1}{4.167} - 0.06 = 0.18$$

This answers the question. ▲

Question 6 *For students entering a college, you are given the following from a multiple decrement model: (1)* 1000 *students enter the college at* $t = 0$. *(2) Students leave the college for failure (1) or all other reasons (2). (3)* $\mu^{(1)}(t) = \mu$, $0 \leq t \leq 4$ *and* $\mu^{(2)}(t) = 0.04$, $0 \leq t < 4$. *(4)* 48 *students are expected to leave the college during their first year due to all causes. Calculate the expected number of students who will leave because of failure during their fourth year.*

Answer It follows from the given information that

$$d_0^{(\tau)} = 1000 \int_0^1 e^{-(\mu + 0.04)t} (\mu + 0.04) \, dt$$

$$= 1000 \left(1 - e^{-(\mu + 0.04)} \right) = 48$$

Hence, $e^{-(\mu + 0.04)} = 0.952$. It follows that

$$\mu + 0.04 = -\ln(0.952) = 0.049$$

and, therefore, that $\mu = 0.009$. This tells us that

$$d_3^{(1)} = 1000 \int_3^4 e^{-0.049t} (0.009) \, dt$$

$$= 1000 \frac{0.009}{0.049} \left(e^{-(0.049)(3)} - e^{-(0.049)(4)} \right) = 7.6$$

This answers the question. ▲

Question 7 *An actuary for an automobile insurance company determines that the distribution of the annual number of claims for an insured chosen at random is modeled by the negative binomial distribution with mean 0.2 and variance 0.4. The number of claims for each individual insured has a Poisson distribution and the means of these Poisson distributions are gamma distributed over the population of those insured. Calculate the variance of this gamma distribution.*

Answer Using the conditional mean and variance formulas, we get

$$E[N] = E_\Lambda (N \mid \Lambda)$$
$$\mathrm{Var}[N] = \mathrm{Var}_\Lambda (E(N \mid \Lambda)) + E_\Lambda (\mathrm{Var}(N \mid \Lambda))$$

Since N, given lambda, is just a Poisson distribution, these equations simplify to

$$E[N] = E_\Lambda (\Lambda)$$
$$\mathrm{Var}[N] = \mathrm{Var}_\Lambda (\Lambda) + E_\Lambda (\Lambda)$$

Using the given values $E[N] = 0.2$ and $\mathrm{Var}[N] = 0.4$, we therefore get

$$0.4 = \mathrm{Var}_\Lambda (\Lambda) + 0.2$$

It follows that $\mathrm{Var}_\Lambda (\Lambda) = 0.2$. ▲

Question 8 *A dam is proposed for a river which is currently used for salmon breeding. You have modeled: (1) For each hour the dam is opened the number of salmon that will pass through and reach the breeding grounds has a distribution with mean 100 and variance 900. (2) The number of eggs released by each salmon has a distribution with mean of 5 and variance of 5. (3) The number of salmon going through the dam each hour it is open and the numbers of eggs released by the salmon are independent. Using the normal approximation for the aggregate number of eggs released, determine the least number of whole hours the dam should be left open so the probability that 10,000 eggs will be released is greater than 95%.*

Answer Let N denote the number of salmon, X the eggs from one salmon, and S the total eggs. Then $E(N) = 100t$ and $\mathrm{Var}(N) = 900t$.

Therefore,

$$E(S) = E(N)E(X) = 500t$$
$$\text{Var}(S) = E(N)\text{Var}(X) + E^2(X)\text{Var}(N)$$
$$= 100t \times 5 + 25 \times 900t = 23{,}000t$$

Hence,

$$P(S > 10{,}000) = P\left(\frac{S - 500t}{\sqrt{23{,}000t}} > \frac{10{,}000 - 500t}{\sqrt{23{,}000t}}\right) = .95$$

Therefore,

$$10{,}000 - 500t = -1.645 \times \sqrt{23{,}!000}\sqrt{t} = -250\sqrt{t}$$
$$40 - 2t = -\sqrt{t}$$
$$2\left(\sqrt{t}\right)^2 - \sqrt{t} - 40 = 0$$
$$\sqrt{t} = \frac{1 \pm \sqrt{1 + 320}}{4} = 4.73$$
$$t = 22.4 \approx 23$$

This answers the question. ▲

Question 9 *For a special fully discrete 20-year endowment insurance on* (55):

 (1)   *Death benefits in year k are given by* $b_k = (21 - k)$, $k = 1, 2, \ldots, 20$.
 (2)   *The maturity benefit is 1.*
 (3)   *Annual benefit premiums are level.*
 (4) $_kV$ *denotes the benefit reserve at the end of year k,* $k = 1, 2, \ldots, 20$.
 (5) $_{10}V = 5.0$.
 (6) $_{19}V = 0.6$.
 (7) $q_{65} = 0.10$.
 (8) $i = 0.08$.

Calculate $_{11}V$.

Answer Let π denote the benefit premium. Since $_{19}V$ is the difference between the actuarial present value of the future benefits and the actuarial present value of the future premiums, we have

$$0.6 = \frac{1}{1.08} - \pi$$

Therefore, $\pi = 0.326$. It follows that

$$_{11}V = \frac{(_{10}V + \pi)(1.08) - (q_{65})(10)}{p_{65}}$$

$$= \frac{(5.0 + 0.326)(1.08) - (0.10)(10)}{1 - 0.10}$$

$$= 5.28$$

This answers the question. ▲

Question 10 *For a stop-loss insurance on a three-person group we are given the following information: (1) Loss amounts are independent. (2) The distribution of loss amount for each person is:*

Loss Amount	Probability
0	0.4
1	0.3
2	0.2
3	0.1

(3) The stop-loss insurance has a deductible of 1 for the group. Calculate the net stop-loss premium.

Answer Let X denote the losses on one life. Then

$$E[X] = (0.3)(1) + (0.2)(2) + (0.1)(3) = 1$$

Now let S denote the total losses. It follows that

$$E[S] = 3E[X] = 3$$
$$E[(S-1)_+] = E[S] - 1(1 - F_s(0))$$
$$= E[S] - (1)(1 - f_s(0))$$
$$= 3 - (1)(1 - 0.4^3)$$
$$= 3 - 0.936$$
$$= 2.064$$

This answers the question. ▲

Question 11 *An insurer's claims follow a compound Poisson claims process with two claims expected per period. Claim amounts can be only 1,2, or 3 and these are equal in probability. Calculate the continuous premium rate that should be charged each period so that the adjustment coefficient will be 0.5.*

Answer We have

$$M_x(r) = E\left[e^{rx}\right] = \frac{e^r + e^{2r} + e^{3r}}{3}$$

$$M_x(0.5) = \frac{e^{0.5} + e + e^{1.5}}{3} = 2.95$$

$$p_1 = E[X] = \frac{1+2+3}{3} = 2$$

$$\lambda[M_x(r) - 1] = cr$$

Since $\lambda = 2$ and $r = 0.5$,

$$2[M_x(0.5) - 1] = 0.5c$$
$$2(2.95 - 1) = 0.5c$$
$$3.9 = 0.5c$$
$$c = 7.8 = \text{ premium rate per period}$$

This answers the question. ▲

Question 12 *For a disability insurance claim, the claimant will receive payments at the rate of 20,000 per year, payable continuously as long as she remains disabled. The length of the payment period in years is a random variable with the gamma distribution with parameters* $\alpha = 2$ *and* $\theta = 1$. *Payments begin immediately and* $\delta = 0.05$. *Calculate the actuarial present value of the disability payments at the time of disability.*

Answer We have

$$\bar{a} = \int_0^\infty \bar{a}_{\overline{t}|} f(t)\, dt = \int_0^\infty \frac{1 - e^{-0.05t}}{0.05} \frac{1}{\Gamma(2)} te^{-t}\, dt$$

$$= \frac{1}{0.05} \int_0^\infty \left(te^{-t} - te^{-1.05t} \right) dt$$

$$= \frac{1}{0.05} \left[-(t+1)e^{-t} + \left(\frac{t}{1.05} + \frac{1}{1.05^2} \right) e^{-1.05t} \right] \Big|_0^\infty$$

$$= \frac{1}{0.05} \left[1 - \left(\frac{1}{1.05} \right)^2 \right] = 1.85941$$

and $20,000 \times .85941 = 37,188.$ ▲

Question 13 *For a special fully discrete 3-year term insurance on* (x), *the level benefit premiums are paid at the beginning of each year,* $i = 0.06$, *and*

k	b_{k+1}	q_{x+k}
0	200,000	0.03
1	150,000	0.06
2	100,000	0.09

Calculate the initial benefit reserve for year 2.

Answer Let π denote the benefit premium and define

$$A = (0.03)(200{,}000)v$$
$$B = (0.97)(0.06)(150{,}000)v^2$$
$$C = (0.97)(0.94)(0.09)(100{,}000)v^3$$

Then the actuarial present value of benefits is

$$A + B + C = 5660.38 + 7769.67 + 6890.08$$
$$= 20{,}320.13$$

Therefore, the actuarial present is

$$\ddot{a}_{x:\overline{3}|}\pi = \left[1 + 0.97v + (0.97)(0.94)v^2\right]\pi$$
$$= 2.7266\pi$$

so that

$$\pi = \frac{20{,}320.13}{2.7266} = 7452.55$$

The initial benefit reserve for Year 2 is therefore

$$_1V + \pi = \frac{(7452.55)(1.06) - (200{,}000)(0.03)}{1 - 0.03} + 7452.55$$
$$= 9411.01$$

This answers the question. ▲

Question 14 *The number of accidents follows a Poisson distribution with mean* 12. *Each accident generates* 1, 2, *or* 3 *claimants with probabilities* $\frac{1}{2}, \frac{1}{3}, \frac{1}{6}$, *respectively. Calculate the variance in the total number of claimants.*

Answer We treat the variances as three independent Poisson variables, corresponding to 1, 2, or 3 claimants.

$$\text{rate}_1 = \tfrac{1}{2} \times 12 = 6 \quad \text{Var}_1 = 6$$
$$\text{rate}_2 = 4 \quad\quad\quad\quad \text{Var}_2 = 4 \times 2^2 = 16$$
$$\text{rate}_3 = 2 \quad\quad\quad\quad \text{Var}_3 = 18$$

Therefore Var $= 6 + 16 + 18 = 40$, due to independence. ▲

Question 15 *For a claims process, you are given: (1) The number of claims* $\{N(t), t \geq 0\}$ *is a nonhomogeneous Poisson process with intensity function*

$$\lambda(t) = \begin{cases} 1 & if \quad 0 \leq t < 1 \\ 2 & if \quad 1 \leq t < 2 \\ 3 & if \quad 2 \leq t \end{cases}$$

(2) Claims amounts Y_i are independently and identically distributed random variables that are also independent of $N(t)$. (3) Each Y_i is uniformly distributed on [200,800]. *(4) The random variable P is the number of claims with claim amount less than 500 by time t = 3. (5) The random variable Q is the number of claims with claim amount greater than 500 by time t = 3. (6) R is the conditional expected value of P, given Q = 4. Calculate R.*

Answer Since

$$\int_0^3 \lambda(t)\, dt = 6$$

it follows that $N(3)$ is Poisson with $\lambda = 6$. Moreover, P is Poisson with mean 3 (since $Prob(Y_i < 500) = 0.5$). Since P and Q are independent, the mean of P is 3, no matter what the value of Q is. ▲

2.14 | The CAS Version of the Actuarial Model Course

In November of 2003, the SOA and CAS versions of the Actuarial Models course took different routes. In addition to the general objective of expecting candidates to be able to apply actuarial models to business applications, as identified in the SOA syllabus the Casualty Actuarial Society placed new emphasis on a list of specific types of models covered in the course:

Survival and Contingent Payment Models

Candidates should be able to work with discrete and continuous univariate probability distributions for failure time random variables. They will be expected to set up and solve equations in terms of life table functions, cumulative distribution functions, survival functions, probability density functions, and hazard functions (e.g., force of mortality), as appropriate. They should have similar facility with models of the joint distribution of two failure times (multiple lives) and the joint distribution of competing risks (multiple decrement). They should be able to formulate and apply stochastic and deterministic models for the present value of a set of future contingent cash flows under an assumed interest rate structure. Candidates also should be able to apply the equivalence principle, and other principles in the text, to associate a cost or pattern of (possibly contingent) costs with a set of future contingent cash flows.

Frequency and Severity Models

Candidates should be able to define frequency (counting) and severity distributions, and be able to use the parameters and moments of these distributions. Candidates also should be able to work with the families of distributions generated by algebraic manipulation and mixing of the basic distributions presented.

Compound Distribution Models

Candidates should be able to calculate the probabilities associated with a compound distribution when the compounding distribution is one of the frequency distributions presented in the syllabus, and the compounded distribution is discrete or a discretization of a continuous distribution. Candidates also should be able to adjust such probability calculations for the impact of policy modifications such as deductibles, policy limits, and coinsurance.

Stochastic Process Models

Candidates should learn to solve problems using stochastic processes. They also should learn how to determine the probabilities and distributions associated with these processes. The following stochastic processes will be covered: Markov chain (discrete-time and continuous-time) processes (see [20], for example), counting processes, Poisson process (including nonhomogeneous and compound Poisson processes), and Brownian motion (see [24]).

Ruin Models

Candidates should be able to analyze the probability of ruin using various models. Other topics covered in this section include the determination of the characteristics of the distribution of the amount of surplus (deficit) at the first time below the initial level and at the lowest level (maximal aggregate loss), and the impact of reinsurance.

Simulation of Models

Candidates should be able to generate discrete and continuous random variables using basic simulation methods. They also should be able to construct algorithms to simulate outcomes using stochastic models.

2.15 | Building Actuarial Models

Whereas Course 3 [2002] dealt with an understanding of actuarial models, the learning objectives of the next actuarial modeling course (formerly Course 4 [2002]) concerned the building of models. According to the Society of Actuaries (see [3]), Course 4 [2002] *"provides an introduction to modeling and covers important actuarial and statistical methods that are useful in modeling. A thorough knowledge of calculus, linear algebra, probability and mathematical statistics is assumed. The candidate will be required to understand the steps involved in the modeling process and how to carry out these steps in solving business problems. The candidate should be able to: (1) analyze data from an application in a business context; (2) determine a suitable model including parameter values; and (3) provide measures of confidence for decisions based upon the model. The candidate will be introduced to a variety of tools for the calibration and evaluation of the models in Course 3 [2002]."*

Ideas and Techniques

In [12], Jones discusses basic types of actuarial models and provides an electronic tool for constructing and studying them. Among the problems discussed are stochastic models, in which given phenomena are represented in probabilistic terms and deterministic ones, where given events are assumed to occur with certainty. He also hints at other types of models, built with relatively new mathematical techniques. This certainly demonstrates the dynamic nature of actuarial science. New problems require new solutions, all the time. As is shown in the book on loss models by Klugman et al. (see [14]), stochastic models include loss models, survival models, contingent payment models, credibility models, linear regression models, stochastic processes, and time-series models. According to the Society of Actuaries syllabus (see [3]), *"the candidate is expected to apply statistical methods to sample data to quantify and evaluate the models presented in Course 3 [2002] and to use the models to solve problems set in a business context."*

Examination Topics

The sample examination consisted of 40 multiple-choice questions. It dealt with the following topics from the SOA and CAS syllabus:

Q1 Invertible autoregressive moving average (ARMA) models, time series, autocorrelation function, MA(1) process, quadratic equations.

Q2 Poisson distributions, means, prior distributions, probability density functions, exponential functions, variances, posterior distributions, factorial function, gamma distributions.

Q3 Auto insurance, randomly selected policies, kurtosis, μ, σ.

Q4 Auto insurance, randomly selected policies, product-limit estimates, survival probabilities, censored data.

Q5 Multiple regression, F-statistic, significant variables, regression coefficients.

Q6 Full credibility standard, expected claims, square-root rule, partial credibility, Bühlmann credibility formula, exposure units.

Q7 Loss models, exponential distributions, maximum likelihood estimates, exponential functions, logarithmic functions, derivatives (calculus).

Q8 Mortalities, reference hazard rates, cumulative relative excess mortalities.

Q9 Dickey-Fuller unit root test, unrestricted regressions, restricted regressions, significance levels, random walk hypothesis, F-distributions, critical values.

Q10 Risk models, claim size distributions, probabilities, independence of claims, Bayesian premiums, variance, expected value.

Q11 Risk models, claim size distributions, probabilities, independence of claims, Bühlmann credibility premiums, expected present value, variance of hypothetical means.

Q12 Random observations, probability density functions, Kolmogorov-Smirnov statistics, integrals.

Q13 Method of least squares, standardized coefficients.

Q14 Mortalities, right-censored data, improper integrals, Aalen estimates, standard deviations, Nelson-Aalen estimators, cumulative hazard functions.

Q15 Mortalities, right-censored data, symmetric confidence intervals, mean survival times, integrals.

Q16 Loss models, Weibull distributions, maximum likelihood estimates, exponential functions, Weibull density functions, logarithmic functions, derivatives (calculus).

Q17 Autoregressive moving average (ARMA)(1,1) models, time series, variance.

Q18 Auto insurance, claim frequencies, Poisson distributions, mean, prior distributions, probability density functions, expected number of claims, exact posterior densities, posterior means, improper integrals.

Q19 Chi-square test, Poisson distributions, means, expected number of observations, exponential functions, factorial function.

Q20 Maximum likelihood estimates, Poisson distributions, negative binomial distributions, negative log likelihoods, likelihood ratio test, null hypothesis.

Q21 Loss models, independence, loss ratios, weighted least squares estimators, derivatives, minima.

Q22 Mortalities, cumulative hazard functions, log-rank test, significance levels, chi-square statistics.

Q23 Risk models, means, variance, expected value, annual process variances, Bühlmann-Straub credibility factor, limits at infinity.

Q24 Claims models, multiple regression, average claim costs, lognormal error components, inflation, linear models, logarithmic functions, exponential functions.

Q25 Loss models, loss ratios, standard deviations, delta method, partial derivatives, variance.

Q26 Auto insurance, random variables, time lag, probabilities, survival functions, right-truncated data.

Q27 Sales models, seasonal adjustments.

Q28 Insurance claims models, expected value, prior probabilities, posterior probabilities.

Q29 Claims models, variance, two-tailed rank-sum hypothesis test, probability distributions, p-values.

Q30 Loss models, exponential distributions, maximum likelihood estimates, deductibles, policy limits, expected payments per loss, inflation, scale parameters, sample means, exponential functions.

Q31 Proportional hazards regressions, Cox models, covariate vectors, partial likelihoods, exponential functions.

Q32 Loss models, nonparametric empirical Bayes credibility premiums, preservation of total losses.

Q33 Annual premium income, loss ratio, two-variable linear regression models, slope coefficients, least-squares estimators, error terms, autoregressive (AR)(1) models, autocorrelation coefficients, standard errors, biased download estimators, Cochrane-Orcutt procedure, consistent estimators of the model slope.

Q34 Automobile insurance, random variable describing the time lag in settling a claim, maximum likelihood estimates, truncated observations, log likelihoods, derivatives (calculus).

Q35 Functions defined by cases, hazard rates, censored claims, kernel-smoothed estimates, bandwidth, biweight kernels, log-transformed confidence intervals.

Q36 Autoregressive moving average (ARMA)(p,q) models, time series, auto-correlation function, simulated series (time series generated by the model), residual of the model, white-noise process, residual autocorrelations, normally distributed random variables, means, variances, Q-statistics, chi-square distributions, degrees of freedom, large displacements.

Q37 Claims models, compensation coverage, Poisson distributions, uniform distributions, posterior probability, posterior distributions, normalizing constants, integrals, posterior density, exponential functions.

Q38 Claims models, compensation coverage, Poisson distributions, uniform distributions, Bühlmann credibility estimates, expected number of claims, variance, Poisson parameter.

Q39 Claims models, independent distributions, exponential distributions, means, standard deviations, second raw moment, component means, quadratic equations.

Q40 Two-variable regression, standard error.

If we examine the frequency of some of the topics and techniques tested in this examination, we come up with the following result: function (26/40), distribution (23/40), mean and standard deviation (14/40), estimate (13/40), variance (12/40), model (11/40), expected value (9/40), probability (9/40), loss (8/40), exponential function (7/40), random variable (7/40), integration (6/40), differentiation (5/40), independence (5/40), insurance (5/40), mortality (5/40). The questions and answers involve various series, formulas, and distributions, such as the Weibull distribution (see [23]), random walks, multiple regression, the chi-square test (see [17]), the gamma distribution (see [23]), confidence intervals, least-squares estimates, and other ideas and techniques. Here are some sample questions that illustrate the "look and feel" of an examination in Course 4:

Examination Questions and Answers

Question 1 *You are given the following information about an invertible ARMA time-series model:*

$$\begin{cases} \rho_1 = -0.4 \\ \rho_k = 0 \qquad k = 2,3,4,\dots \end{cases}$$

Determine θ_1.

Answer Because the autocorrelation function is zero starting with lag 2, this must be an $MA(1)$ model. Then

$$-.4 = \rho_1 = \frac{-\theta_1}{1 + \theta_1^2}$$

so that

$$-.4 - .40_1^2 = -\theta_1$$

Therefore,

$$.40_1^2 - \theta_1 + .4 = 0$$

This quadratic equation has two roots, 0.5 and 2. Because the coefficient's absolute value must be less than 1, only 0.5 is acceptable. ▲

Question 2 *You are given that the full credibility standard is* 100 *expected claims and that the square-root rule is used for partial credibility. You approximate the partial credibility formula with a Bühlmann credibility formula by selecting a Bühlmann k value that matches the partial credibility formula when* 25 *claims are expected. Determine the credibility factor for the Bühlmann credibility formula when* 100 *claims are expected.*

Answer The number of expected claims (e) is proportional to the number of exposure units (n). Let $e = cn$. Using Bühlmann credibility and partial credibility gives

$$\sqrt{\frac{25}{100}} = \frac{1}{2} = \frac{25/c}{25/c + k} = \frac{25}{25 + ck}$$

Therefore, $ck = 25$. When we have 100 expected claims,

$$Z = \frac{100/c}{100/c + k} = \frac{100}{100 + ck} = \frac{100}{100 + 25} = 0.80$$

This answers the question. ▲

Question 3 *A Dickey-Fuller unit root test was performed on* 100 *observations of each of three price series by estimating the unrestricted regression*

$$Y_t - Y_{t-1} = \alpha + \beta t + (\rho - 1) Y_{t-1}$$

and then the restricted regression

$$Y_t - Y_{t-1} = \alpha$$

You are given that

Price Series	Unrestricted Error Sums	Restricted Error Sums
I	3233.8	3552.2
II	1131.8	1300.5
III	211.1	237.0

*and that the critical value at the 0.01 significance level for the F-distribution cal-
culated by Dickey and Fuller is 5.47. For which series do you reject at the 0.10
significance level the hypothesis of a random walk?*

Answer We know that the F statistics are given by the formula

$$F = (N-k)\frac{(ESS_R - ESS_{UR})}{q(ESS_{UR})}$$

and $N = 100$, $k = 3$, $q = 2$. Therefore,

Series I: $F = 97\dfrac{(3552.2 - 3233.8)}{2(3233.8)} = 4.78$ (Fail to reject)

Series II: $F = 97\dfrac{(1300.5 - 1131.8)}{2(1131.8)} = 7.23$ (Reject)

Series III: $F = 97\dfrac{(237.0 - 211.1)}{2(211.1)} = 5.95$ (Reject)

This answers the question. ▲

Question 4 *For a mortality study with right-censored data, you are given:*

t_i	d_i	Y_i	$\frac{d_i}{Y_i(Y_i - d_i)}$	$\widehat{S}(t_i)$	$\int_{t_i}^{\infty}\widehat{S}(t)\,dt$
1	15	100	0.0018	0.8500	14.424
8	20	65	0.0068	0.5885	8.474
17	13	40	0.0120	0.3972	3.178
25	31	31	—	0.0000	0.000

Determine the symmetric 95% confidence interval for the mean survival time.

Answer By the definition of $\widehat{\mu}_\tau$ we have

$$\widehat{\mu}_\tau = \int_0^\tau \widehat{S}(t)\,dt$$

$$= (1.0 \times 1) + (0.85 \times 7) + (0.5885 \times 9) + (0.3972 \times 8) = 15.42$$

Therefore,

$$\hat{V}[\hat{\mu}_\tau] = \sum_{i=1}^{D} \left[\int_{t_i}^{\tau} \hat{S}(t)\, dt \right]^2 \frac{d_i}{Y_i(Y_i - d_i)}$$

$$= 14.424^2 \times 0.0018 + 8.474^2 \times 0.0068 + 3.178^2 \times 0.0120$$

$$= 0.9840$$

We conclude that the 95% confidence interval is $15.42 \pm 1.96 \times \sqrt{0.9840}$. ▲

Question 5 *A sample of ten losses has the following statistics:*

$$\sum_{i=1}^{10} X^{-2} = 0.00033674 \qquad \sum_{j=1}^{10} X^{0.5} = 488.97$$

$$\sum_{i=1}^{10} X^{-1} = 0.023999 \qquad \sum_{j=1}^{10} X = 31,939$$

$$\sum_{i=1}^{10} X^{-0.5} = 0.34445 \qquad \sum_{j=1}^{10} X^2 = 211,498,983$$

You assume that the losses come from a Weibull distribution with $\tau = 05$. Determine the maximum likelihood estimate of the Weibull parameter θ.

Answer The Weibull density function is $f(x) = .5\,(x\theta)^{-.5}\, e^{-(x/\theta)^{.5}}$. Therefore the likelihood function is

$$L(\theta) = \prod_{j=1}^{10} .5\,(x_j\theta)^{-.5}\, e^{-(x_j/\theta)^{.5}}$$

$$= (.5)^{10} \left(\prod_{j=1}^{10} x_j \right)^{-.5} \theta^{-5} e^{-\theta^{-.5} \sum_{j=1}^{10} x_j^{.5}}$$

$$\propto \theta^{-5} e^{-488.97\theta^{-.5}}$$

The logarithm and its derivative are

$$l(\theta) = -5\ln\theta - 488.97\theta^{-.5}$$

$$l'(\theta) = -5\theta^{-1} + 244.485\theta^{-1.5}$$

Setting the derivative equal to zero yields

$$\hat{\theta} = (244.485/5)^2 = 2391$$

This answers the question. ▲

Question 6 *You are using an ARMA(1,1) model to represent a time series of* 100 *observations. You have determined:*

$$\begin{cases} \hat{y}_{100}(1) & = \quad 197.0 \\ \hat{\sigma}_{\varepsilon}^2 & = \quad 1.0 \end{cases}$$

Later, you observe that y_{101} *is* 188.0. *Determine the updated estimate* σ_{ε}^2.

Answer The estimated variance of the forecast errors is the sum of the squares of the error terms divided by $T - p - q$. In this case, after 100 observations, the sum of the squares of the error terms must equal 98, because the sum divided by $(100 - 1 - 1)$, or 98, is 1.0.

The 101st observation introduces a new error term equal to $188 - 197 = -9$. The square of this term is 81. Adding 81 to the previous sum of 98 gives a new total of 179. Dividing 179 by $101 - 1 - 1 = 99$ gives a new estimated variance of 1.8. ▲

Question 7 *You are given: (1) An individual automobile insured has annual claim frequencies that follow a Poisson distribution with mean* λ. *(2) An actuary's prior distribution for the parameter* λ *has probability density function*

$$\pi(\lambda) = (0.5)\, 5e^{-5\lambda} + (0.5)\frac{1}{5}e^{-\lambda/5}$$

(3) In the first policy year, no claims were observed for the insured. Determine the expected number of claims in the second policy year.

Answer The posterior distribution is

$$\pi(\lambda \mid 0) \propto e^{-\lambda}\left[(.5)\,5e^{-5\lambda} + (.5).2e^{-.2\lambda}\right]$$
$$= 2.5e^{-6\lambda} + .1e^{-1.2\lambda}$$

The normalizing constant can be obtained from

$$\int_0^{\infty} \left[2.5e^{-6\lambda} + .1e^{-1.2\lambda}\right] d\lambda = .5$$

and therefore the exact posterior density is $\pi(\lambda \mid 0) = 5e^{-6\lambda} + .2e^{-1.2\lambda}$.

The expected number of claims in the next year is the posterior mean,

$$E(\Lambda \mid 0) = \int_0^{\infty} \lambda \left[5e^{-6\lambda} + .2e^{-1.2\lambda}\right] d\lambda$$
$$= \frac{5}{36} + \frac{5}{36} = \frac{5}{18} = .278$$

This answers the question. ▲

Question 8 *During a one-year period, the number of accidents per day was distributed as follows:*

Accidents	0	1	2	3	4	5
Days	209	111	33	7	3	2

You use a chi-square test to measure the fit of a Poisson distribution with mean 0.60. The minimum expected number of observations in any group should be 5. The maximum possible number of groups should be used. Determine the chi-square statistic.

Answer There are 365 observations, so the expected count for k accidents is

$$365 p_k = 365 \frac{e^{-.6}(.6)^k}{k!}$$

which produces the following table:

Accidents	Observed	Expected	Chi-square
0	209	200.32	0.38
1	111	120.19	0.70
2	33	36.06	0.26
3	7	7.21	1.51
4	3	1.08	
5	2	0.14	

This answers the question. ▲

Question 9 *Twenty independent loss ratios $Y_1, Y_2, \ldots, Y_{20}$ are described by the model*

$$Y_t = \alpha + \varepsilon_t$$

where

$$Var(\varepsilon_t) = 0.4, \quad t = 1, 2, \ldots, 8$$
$$Var(\varepsilon_t) = 0.6, \quad t = 9, 10, \ldots, 20$$

You are given:

$$\overline{Y}_1 = \frac{1}{8}(Y_1 + Y_2 + \cdots + Y_8)$$
$$\overline{Y}_2 = \frac{1}{12}(Y_9 + Y_{10} + \cdots + Y_{20})$$

Determine the weighted least squares estimator of α in terms of $\overline{Y}_1$ and $\overline{Y}_2$.

Answer We need

$$S(\alpha) = \sum_{t=1}^{8} \left(\frac{Y_t - \alpha}{\sqrt{0.4}} \right)^2 + \sum_{t=9}^{20} \left(\frac{Y_t - \alpha}{\sqrt{0.6}} \right)^2$$

to be a minimum. Setting the derivative equal to zero produces the equation

$$S'(\alpha) = \frac{1}{.4} \sum_{t=1}^{8} 2(Y_t - \alpha) + \frac{1}{.6} \sum_{t=9}^{20} 2(Y_t - \alpha) = 0$$

Multiplying by 0.6 produces the equation

$$0 = 3\left(8\bar{Y}_1 - 8\alpha\right) + 2\left(12\bar{Y}_2 - 12\alpha\right)$$

$$0 = 24\bar{Y}_1 + 24\bar{Y}_2 - 48\alpha$$

$$\alpha = .5\bar{Y}_1 + .5\bar{Y}_2$$

This answers the question. ▲

Question 10 *For a mortality study, you are given: (1) Ten adults were observed beginning at age 50. (2) Four deaths were recorded during the study at ages 52, 55, 58, and 60. The six survivors exited the study at age 60. (3) H_0 is a hypothesized cumulative hazard function with values as follows:*

$H_0(50) = 0.270$	$H_0(51) = 0.280$	$H_0(52) = 0.290$	$H_0(53) = 0.310$
$H_0(54) = 0.330$	$H_0(55) = 0.350$	$H_0(56) = 0.370$	$H_0(57) = 0.390$
$H_0(58) = 0.410$	$H_0(59) = 0.435$	$H_0(60) = 0.465$	

Determine the result of the one-sample log-rank test used to test whether the true cumulative hazard function differs from H_0. The possible answers are

(A) *Reject at the 0.005 significance level.*

(B) *Reject at the 0.01 significance level, but not at the 0.005 level.*

(C) *Reject at the 0.025 significance level, but not at the 0.01 level.*

(D) *Reject at the 0.05 significance level, but not at the 0.025 level.*

(E) *Do not reject at the 0.05 significance level.*

Answer We are given that $O = 4$ and

$$E = (.29 - .27) + (.35 - .27) + (.41 - .27) + 7(.465 - .27)$$
$$= 1.605$$

Therefore, the chi-square statistic is

$$(4 - 1.605)^2 / 1.605 = 3.57$$

Hence the 0.05 level of significance is 3.84. So the answer is (E). ▲

Question 11 *Your claims manager has asserted that a procedural change in the claims department implemented on January 1, 1997 immediately reduced claim severity by 20 percent. You use a multiple regression model to test this assertion. For the dependent variable, Y, you calculate the average claim costs on closed claims by year during 1990–1999. You define the variable X as the year. You also define a variable D as:*

$$D = \begin{cases} 0 & \text{for years 1996 and prior} \\ 1 & \text{for years 1997 and later} \end{cases}$$

Assuming a lognormal error component and constant inflation over the entire period, which of the following models would be used to test the assertion? The possible answers are

- (A) $Y = \alpha_1^D \beta_1^X \varepsilon$
- (B) $Y = \alpha_1 \alpha_2^D \beta_1^X \varepsilon$
- (C) $Y = \alpha_1 \beta_1^X \beta_2^{XD} \varepsilon$
- (D) $Y = \alpha_1 \alpha_2^X \beta_1^X \beta_2^{XD} \varepsilon$
- (E) $Y = \alpha_1 \alpha_2^D X^{\beta_1} \varepsilon$

Answer With a lognormal error component, the linear model should be for the logarithm of the observation. A model that conforms to the description is

$$\ln Y = \alpha_1^* + \alpha_2^* D + \beta_1^* X + \varepsilon^*$$

Exponentiating both sides yields

$$Y = e^{\alpha_1^*} e^{\alpha_2^* D} e^{\beta_1^* X} e^{\varepsilon^*}$$

and then defining an unstarred quantity as its starred version exponentiated, we have

$$Y = \alpha_1 \alpha_2^D \beta_1^X \varepsilon$$

Note that when D is 1, the value of Y is multiplied by α_2 and so the hypothesis to test is if this value is equal to 0.8. ▲

Question 12 *Two eight-sided dice, A and B, are used to determine the number of claims for an insured. The faces of each die are marked with either 0 or 1, representing the number of claims for that insured for the year.*

Die	Pr(Claims=0)	Pr(Claims=1)
A	$\frac{1}{4}$	$\frac{3}{4}$
B	$\frac{3}{4}$	$\frac{1}{4}$

Two spinners, X and Y, are used to determine claim cost. Spinner X has two areas marked 12 and c. Spinner Y has only one area marked 12.

Spinner	Pr(Cost=12)	Pr(Cost=c)
X	$\frac{1}{2}$	$\frac{1}{2}$
Y	1	0

To determine the losses for the year, a die is randomly selected from A and B and rolled. If a claim occurs, a spinner is randomly selected from X and Y and spun. For subsequent years, the same die and spinner are used to determine losses. Losses for the first year are 12. Based upon the results of the first year, you determine that the expected losses for the second year are 10. Calculate c.

Answer Let *EV* stand for "expected value," *priorProb* for "prior probability," and *postProb* for "posterior probability."
Then

Die/Spinner	*priorProb*	**Probability of getting a 12**	*postProb*
AX	$\frac{1}{4}$	$\frac{3}{4} \times \frac{1}{2} = \frac{3}{8}$	$\frac{1}{4}$
AY	$\frac{1}{4}$	$\frac{3}{4} \times 1 = \frac{3}{4}$	$\frac{1}{2}$
BX	$\frac{1}{4}$	$\frac{1}{4} \times \frac{1}{2} = \frac{1}{8}$	$\frac{1}{12}$
BY	$\frac{1}{4}$	$\frac{1}{4} \times 1 = \frac{1}{4}$	$\frac{1}{6}$
Total	1	$\frac{3}{2}$	1

and

Die/Spinner	*EV*	*postProb*	*EV* × *postProb*
AX	$\frac{3}{4} \times \frac{1}{2} \times (12+c)$	$\frac{1}{4}$	$1.125 + \frac{3}{32}c$
AY	$\frac{3}{4} \times (12)$	$\frac{1}{2}$	4.5
BX	$\frac{1}{4} \times \frac{1}{2} \times (12+c)$	$\frac{1}{12}$	$0.125 + \frac{1}{96}c$
BY	$\frac{1}{4} \times (12)$	$\frac{1}{6}$	0.5
Total		1	$6.25 + \frac{10}{96}c$

Since the expected value is 10 and $6.25 + \frac{10}{96}c = 10$, we have $c = 36$. ▲

Question 13 *For a study in which you are performing a proportional hazards regression using the Cox model, you are given that*

$$h(t|Z) = h_0(t)\exp\left(\beta'Z\right)$$

and that the covariate vectors for the three individuals studied, in the order in which they die, are as follows:

$$Z_1 = \begin{pmatrix} 1 \\ 0 \\ 0 \end{pmatrix} \quad Z_2 = \begin{pmatrix} 0 \\ 1 \\ 0 \end{pmatrix} \quad Z_3 = \begin{pmatrix} 0 \\ 0 \\ 1 \end{pmatrix}$$

Determine the partial likelihood.

Answer By definition, we have

$$L(\hat{a}) = \left(\frac{e^{\beta_1}}{e^{\beta_1} + e^{\beta_2} + e^{\beta_3}} \right) \left(\frac{e^{\beta_2}}{e^{\beta_2} + e^{\beta_3}} \right) \left(\frac{e^{\beta_3}}{e^{\beta_3}} \right)$$

This answers the question. ▲

Question 14 *You are given the following claims settlement activity for a book of automobile claims as of the end of 1999:*

Year Reported/Year Settled	1997	1998	1999
1997	Unknown	3	1
1998		5	2
1999			4

and

$$L = (\textit{Year Settled} - \textit{Year Reported})$$

is a random variable describing the time lag in settling a claim. The probability function of L is

$$f_L(l) = (1 - p) p^l, \textit{ for } l = 0, 1, 2, \ldots$$

Determine the maximum likelihood estimate of the parameter p.

Answer The observations are right and left truncated and the truncation depends upon the report year. For report year 1997 only claims settled at durations 1 and 2 can be observed, so the denominator must be the sum of those two probabilities. For 1998, only durations 0 and 1 can be observed and for 1999 only duration 0 can be observed. Calculation of the denominator probabilities is summarized below.

Probabilities

Year Reported	Settled in 1998	Settled in 1999	Sum (Denominator)
1997	$(1-p)p$	$(1-p)p^2$	$(1-p)p(1+p)$
1998	$(1-p)$	$(1-p)p$	$(1-p)(1+p)$
1999		$(1-p)$	$(1-p)$

The likelihood function is

$$L(p) = ABCDE = \frac{p^3}{(1+p)^{11}}$$

where

▶ $A = \left(\dfrac{(1-p)p}{(1-p)p(1+p)} \right)^3 = \left(\dfrac{1}{1+p} \right)^3$

▶ $B = \left(\dfrac{(1-p)p^2}{(1-p)p(1+p)} \right)^1 = \left(\dfrac{p}{1+p} \right)^1$

▶ $C = \left(\dfrac{(1-p)}{(1-p)(1+p)} \right)^5 = \left(\dfrac{1}{1+p} \right)^5$

▶ $D = \left(\dfrac{(1-p)p}{(1-p)(1+p)} \right)^2 = \left(\dfrac{p}{1+p} \right)^2$

▶ $E = \left(\dfrac{1-p}{1-p} \right)^4 = 1$

The log likelihood is therefore

$$l(p) = 3\ln(p) - 11\ln(1+p)$$

Taking the derivative with respect to p, we obtain the equation to solve:

$$\frac{3}{p} - \frac{11}{(1+p)} = 0$$

Therefore, the solution is $\widehat{p} = \frac{3}{8}$. ▲

Question 15 *For a two-variable regression based on seven observations, you are given:*

(1) $\sum (X_i - \overline{X})^2 = 2000$

(2) $\sum \widehat{\varepsilon}_i^2 = 967$

Calculate $s_{\widehat{\beta}}$, the standard error of $\widehat{\beta}$.

Answer Since

$$s^2 = \frac{\sum \widehat{\varepsilon}_i^2}{N-2} = \frac{967}{5} = 193.4$$

$$s_{\widehat{\beta}}^2 = \frac{s^2}{\sum x_i^2} = \frac{193.4}{2000} = 0.0967$$

it follows that $s_{\widehat{\beta}} = 0.31$. ▲

2.16 | Advanced SOA Examination Topics

Becoming a Fellow of the Society of Actuaries is similar to becoming a member of other professions such as doctors and lawyers. The current accreditation process was described earlier in the book. The process is arduous and involves several components. Passing Exam P, Exam FM, Exam M (MFE or MLC), and Exam C are the first formal steps toward becoming an actuary. If you have passed these examinations, as well as the VEE, FAP, and APC components described above, you become an *Associate* of the Society of Actuaries and are entitled to put the title ASA after your name.

For many members of the Society, becoming an Associate is also the last step because the work they do does not require them to pass additional examinations. The term "Career Associates" has been coined informally for those who stop writing examinations at this point in their careers.

The following specialized courses referred to in the survey illustrate the advanced topics and techniques that make up the toolbox of full-fledged actuaries. These courses are described in past versions of the Basic Education Catalogs of the Society of Actuaries. While the organization of these advanced courses has changed in the past few years since the survey, much of the same material is covered and thus the topics here are still very relevant.

SOA Associateship Course 5 [2002]—Application of Basic Actuarial Principles

According to an earlier SOA syllabus, this course "develops the candidate's knowledge of basic actuarial principles applicable to a variety of financial security systems: life, health, property and casualty insurance, annuities, and retirement systems. The candidate will be required to understand the purpose of these systems, the design and development of financial security products, the concepts of anti-selection and risk classification factors, and the effects of regulation and taxation on these issues. The course will develop the candidate's knowledge of principles and practices applicable to the determination of premiums and rates and the valuation and funding of these financial security systems." The topics covered in this course were divided into the following topic areas:

- *Basic Principles of Design.* Here it is expected that you can explain and deal with problems of financial insecurity, product development, and methods of distribution.

- *Basic Principles of Risk Classification.* Here you are expected to classify risks involved in life insurance, health insurance, retirement plans, property and casualty insurance, and in nontraditional areas of insurance such as warranty. You are also expected to be able to evaluate the risk classification factors and be able to carry out a cost/benefit analysis.

- *Basic Principles of Pricing/Ratemaking/Funding.* Here you are expected to be able to describe the objectives of various coverages, evaluate the assumptions underlying pricing, and describe the major pricing and funding techniques and methods used in life insurance, health insurance, retirement plans, and property and casualty insurance. You're also expected to be able to develop different types of profit/surplus measure and describe methods for evaluating pricing.

- *Basic Principles of Valuation.* Here you are expected to be able to describe valuations and the different purposes for performing a valuation, and be able to determine the actuarial value resulting from applying the methodology. You are also expected to be able to interpret the results of the valuation.

SOA Fellowship Course 6 [2002]—Finance and Investment

As stated in an earlier SOA syllabus, this course "extends the candidate's knowledge of basic actuarial principles in the fields of investments and asset management. Candidates completing this course will have developed some expertise in the areas of capital markets, investment vehicles, derivatives-applications, principles of portfolio management and asset-liability management."

SOA Fellowship Course 7 [2002]—Applied Modeling

According to an earlier SOA syllabus, this course, for which laptops were required, "introduces the candidate to the practical considerations of modeling through an intensive seminar using a case study format. The interactive approach of the seminar will require candidates to draw upon knowledge from the basic courses and learn applied modeling skills in a hands-on environment. The seminar also emphasizes communication skills, teamwork and the synthesis of subjects in an applied setting."

Course 7 drew on students' experience in modeling, problem solving, and communication and involved sufficient technical knowledge of a limited number of models to be able to benefit from the course. It was expected that students were familiar with the idea of an actuarial model and had a broad understanding of their use in actuarial practice involving survival models, credibility models, risk theory models, ruin theory models, option pricing models, cash flow and cash flow testing models, and nontraditional models. Students needed to be able to apply appropriate models to solve business problems and had to be able to analyze and understand the results of the modeling process. Moreover, they needed to be able to effectively explain their work and results to others.

SOA Fellowship Course 8 [2002]—Advanced Specialized Actuarial Practice

This course was divided into several options: finance (corporate, capital management, financial risk management, financial strategies); health, group life, and managed care (plan design, data and cost analysis and rating, financial management, administration and delivery systems); individual insurance (marketing of individual life insurance and annuity products, pricing, valuation and financial statements, product development and design); investments (portfolio management, option pricing techniques, asset-liability management); and retirement benefits.

2.17 | CAS Courses Referenced in the Survey

As in the SOA case, passing the beginning CAS courses were and are the stepping stones to becoming an actuary. You can then use the letters ACAS after your name.

Prior to the update to the CAS examination structure, after passing the remaining examinations in CAS Courses 6 [2003] through 9 [2003], Associates became *Fellows* and used the letters FCAS. Starting in 2003, the CAS version of Course 3 [2003] became different from that of the SOA version of Course 3 [2002] and candidates had to write a CAS-only examination in Course 3 [2003]. The syllabus of Courses 5 [2003] through 9 [2003] was based on the work experience of the candidates.

For many members of the Society, becoming an Associate was and is the last step because the work they do does not require them to pass additional examinations.

Since the actual course content of the advanced CAS courses is now organized differently, the courses described below are included to give you a global sense of how the profession has evolved over the last ten years.

CAS Associateship Course 5 [2003]—Introduction to Property and Casualty Insurance and Ratemaking

The CAS Course 5 [2003] mentioned in the survey dealt in part with the legal and commercial nature of insurance policies and coverage. Actuaries should be able to understand the fine print on an insurance policy. They should have "an understanding of the nature of the coverages provided and the exposure bases used in the respective lines of insurance." They should understand the connection between coverage and pricing and be able to interpret the conditions, exclusions, and limitations of P/C policies. To do so, they must be familiar with manual excerpts and must study illustrative parts of relevant manuals dealing with forms, coverages, and rating process.

The course also covered insurance company operations including company organization, marketing and distributions systems, underwriting, and claims. In addition, P/C actuaries need to have a thorough understanding of the underwriting function including purpose, principles, and activities. They should also know how to settle claims based on policy provisions and have an understanding of the impact of settlements on overall loss levels.

P/C actuaries must understand the basic principles of ratemaking and be able to analyze data, select appropriate techniques, and have the tools to solve numerical problems. They should be able to compare the relative advantages and disadvantages of different procedures. In more general terms, P/C actuaries must be able to relate changes in the economic environment to the pricing of insurance.

CAS Fellowship Course 6 [2003]—Reserving, Insurance Accounting Principles, and Reinsurance

The components of CAS Course 6 [2003] mentioned in the survey were statements of principles and standards of practice of insurance, dynamic financial analysis (DFA), expense analysis, published financial information, and reinsurance. In particular, actuaries should be able to establish and review actuarial reserves, and select and evaluate loss reserving methods for known claims and for claims incurred but not yet reported (IBNR). Property and casualty actuaries must have a general knowledge of insurance accounting. They must be able to explain the differences between the different accounting methods and be able to interpret and evaluate numerical data from the reports. In addition, they must understand the ideas and techniques involved in insurance companies insuring other insurance companies (an activity known as "reinsurance"). They must be familiar with different types of reinsurance, the purposes of reinsurance, and how it is marketed and underwritten. They must understand how concepts such as pricing and reserving are adapted to apply to reinsurance.

CAS Fellowship Course 7 [2003]—Annual Statement, Taxation and Regulation (Nation-Specific)

- **(United States)**

  The U.S. version of CAS Course 7 [2003] mentioned in the survey was country-specific since it involved reporting principles, taxation, and other regulations that differ from country to country. The course was offered in two flavors, American and Canadian. It consisted of two main parts: insurance law and regulations, and accounting. Property and casualty actuaries must understand different aspects of insurance regulation and laws, markets, coverages, and private and governmental programs. They must be able to assess how these regulations and laws impact on property/casualty coverages, ratemaking, and pricing. In the United States, this includes tort law, statutory insurance, and governmental programs such as social security and Medicare, catastrophes, and workers' compensation. They must also understand the role of antitrust law as it pertains to insurance regulation. They must understand the impact of government regulations on ratemaking, profitability, risk classification, and the availability of insurance. The course also covers the regulation for solvency, the IRIS [Insurance Regulatory Information Systems] financial ratio test, and guaranty fund mechanisms set up by the various states. The U.S. version of the course covers the aspects of statutory and GAAP [generally accepted accounting principles] insurance accounting and taxation as they affect reserving and statutory reporting, and insurance company audits. The course assumed a working knowledge of general accounting such as that gained from Course 6.

- **(Canada)**

  The Canadian version of CAS Course 7 [2003] mentioned in the survey included a comprehensive presentation of Canadian tort law in the perspective of the insurance business in Canada. The course focused on insurance regulation and insurance contract law and included an overview of federal and provincial insurance programs. It also covered finance and solvency issues. It included insurance accounting and its relevant laws and regulations. The course also included sections on background law and insurance, the regulations of insurance in Canada, insurance as an essential service, and federal and provincial government plans such as the principles and ideas underlying Canadian employment insurance and the Canadian pension programs. The course included material regarding environmental liabilities in the United States and on Canadian earthquake guidelines. It also covered Canadian provincial health plans and the regulatory environment surrounding U.S. workers' compensation. Canadian P/C actuaries must also understand Canadian automobile insurance programs including no-fault concepts and residual market requirements, and provincial guaranty funds. The finance and solvency section of the course dealt with finance, taxation, and solvency tests.

Canadian P/C actuaries need to be familiar with the concept of an *Annual Return*. This topic therefore included guidelines from OSFI [Office of the Superintendent of Financial Institutions] and the provincial regulatory bodies. A thorough knowledge of GAAP [generally accepted accounting principles] was also required. The course covered solvency monitoring systems such as the minimum capital test, risk-based capital requirements, and the DCAT [dynamic capital adequacy testing] method of the Canadian Institute of Actuaries.

CAS Fellowship Course 8 [2003]—Investments and Financial Analysis

The CAS Course 8 [2003] mentioned in the survey dealt with a broad array of finance, investment, and financial risk management topics. Its two main parts were financial theory and financial analysis. The course built on the topics covered in Course 2 [2003]. It also assumed knowledge about liability and reserve risk from Course 6 [2003], some knowledge of underwriting from Course 5 [2003], and knowledge of models and modeling from Courses 3 [2003] and 4 [2003].

Financial theory dealt with investments with an emphasis on the cash flow characteristics, value, and risks inherent in various financial instruments. In particular, it dealt with financial instruments and markets, portfolio theory, equilibrium in capital markets, CAPM [capital asset pricing model], index models, and arbitrage pricing. One of the key concepts covered was that of market efficiency. In addition, the financial theory part of the course covered fixed income securities, options, futures, and swaps, and international securities.

The financial analysis part of the course emphasized measuring and managing the financial risk and overall value of an insurance company. It included asset liability management and factors that affect the price sensitivity of fixed income securities. It also dealt with various ways in which a portfolio manager can manage the interest rate and cash flow risk in a portfolio.

CAS Fellowship Course 9 [2003]—Advanced Ratemaking, Rate of Return, and Individual Risk Rating Plans

CAS Course 9 [2003] dealt with "the types of practical problems that a fully qualified actuary working in ratemaking should be able to solve." The techniques covered were divided into four sections: classification ratemaking topics; excess and deductible rating; rate of return; and the loading for risk. The excess and deductible section dealt with methods of estimating losses within layers of coverage. The rate-of-return part of the course *explored the relationship between insurance concepts (such as underwriting profits, premium-to-surplus ratios, and investment income) and financial concepts (such as interest rates, inflation rates, cost of capital, and risk premiums).* The loading-for-risk part of the course concentrated on the fortuitous nature of insurance claims and the fact that the loading for profit in rates may not be realized. Individual risk rating is one of the important functions performed by an actuary in the rating of individual risks. The earlier courses dealt mainly with group insurance and classification risk rating. Course 9 [2003], on the other hand, was meant to enable P/C actuaries to design and manage individual risk rating systems. The three key concepts involved are experience rating (using individual risk experience to adjust rates), retrospective rating (using individual risk experience to adjust premiums after the completion of policies), and excess and deductible rating (excluding portions of the individual risk experience from insurance coverage, and prospectively reducing rates). It was assumed that students *had a good working knowledge of credibility, loss limitation, and rate modification concepts as they apply to prospective and retrospective rating. In addition, they will be expected to have knowledge of loss distribution, insurance charge, and excess loss charge concepts as they apply to loss retention programs.*

2.18 | Professional Development Course

As is the case for other professions such as engineering, accounting, law, dentistry, and medicine, and so on, members of the actuarial profession are bound by explicit rules of conduct and a profession-specific code of ethics. Actuarial education programs usually include a professional development component. In all accreditation systems, be they university-based, profession-based, or a hybrid of the two, the professional societies of the respective countries examine candidates in this area before admitting them to the profession. The specific rules for accreditation are country-specific and are explained in detail on the websites listed at the end of the book.

Chapter 3

ACTUARIAL JOBS

Actuaries primarily provide actuarial services. Given their role in society world-wide, it is not surprising that a web search under *actuarial services* may produce over 200,000 hits. Any such search shows that actuarial employers come in all sizes, large and small. Actuaries work not only in consulting firms and insurance companies, they also work in government departments, in the human resource offices of most major companies, and in small firms specializing in a variety of actuarial tasks.

It is difficult to present an accurate snapshot of the state of the actuarial world at any given moment in time. Globalization of economies and changes in finan-cial regulations and structures often blur the distinction between the actuarial and non-actuarial role of many companies. In this text, therefore, we concentrate on representative companies, large and diversified enough to illustrate the spectrum of actuarial careers. You will find basic information about some of the top employ-ers of actuaries. The list does not include alternative forms of employment in business, banking, teaching, human resource management, administration, gov-ernment, and so on. For more detailed information on actuarial careers elsewhere, we refer you to the websites listed in Appendix D at the end of this book. The profiles vary from company to company to illustrate different aspects of actuarial employment.

As is the case in many professions, social networking has become a signif-icant component of job search and professional advancement. We illustrate the role of social networking in the lives of actuaries by including some selected but somewhat random search results. This will give you an idea of how you might go about developing your own professional social network, both as an actuarial student and as an established actuary. For this reason, we sometimes use excerpts from Facebook, LinkedIn, and Twitter to illustrate the usefulness of social media to actuarial career searches.

We begin by answering some of the basic questions you may have about getting started with your actuarial career.

3.1   Landing Your First Job

Q **What advice and tips would you give, from an employer's point of view, to students applying for an actuarial position in your company, based on your participation in recruiting activities?**

Answer   Do more "not-actuarial-related activities," be yourself in interviews, don't try to do too much in interviews, show you are more than just a worker, but also a great person, etc. (There is more to life than grades and exams!)

Answer   Write exams as quickly as possible.

Answer   Be honest. Be yourself. Good grades are important, but companies look more and more for extracurricular activities. It's better to have a B but be involved in your actuarial association while doing volunteer work for a youth center than devoting your life to your A+. They want to know you're going to be efficient but also fun to work with, that you'll have something different to bring the company. Participate in conventions, wine and cheese events, and meetings with future employers. Even if you don't give them your CV, they'll probably remember you.

Answer   To be energetic, to have a good team spirit, to be able to have activities while studying. Internship work is also well recognized. Good grades (not necessarily excellent grades) are also required.

Answer   Have at least two exams (for full-time). Have someone proofread your letter and résumé, or at least use spell check—people who misspell "actuarial" are eliminated. I should have the idea that you are writing to me, or my company. State that you know this is a casualty company, and use CAS exams (not SOA). If you are a really good candidate (exams, A+ marks in an actuarial program, work experience whether in insurance or not, a great personality), you don't need to apply to a vast number of companies to get a job—you have time to customize. If you aren't a really good candidate, then you have to customize to get my attention. Keep the cover letter to one page. Take interview training. Be able to answer the standard questions. The more you talk, the more I will feel that I know you when you leave the interview. Look at our website before you apply to us (or for interns, before your interview)—almost everyone does. Work at something—it gives of us more to talk about.

Answer Graduate from university with as many exams as possible. Get relevant work experience through internships and summer jobs. Actuarial employers are generally located in large cities so be prepared to move to one if you don't already live in one. Make sure you develop both actuarial, computer, and communications skills. Pick the job that you'll think you'll enjoy the most—don't necessarily pick the one offering the highest salary.

Answer Always apply for job openings. Don't be afraid of a company's reputation. Grades are not everything.

Answer Take the internship opportunities to taste various forms of work and working environments. Try to see where you feel better. Then come to see me and tell me why you want to come and work for me.

Answer Résumé must be free of grammatical errors. Good marks are the most important factor—they need not all be 90+, but we do not interview very many students with an average below 70. Come to the interview well-dressed and professional. Look like you would be ready to work today. If in doubt about "business casual," take it up a notch. Listen to what the interviewer is saying. You will be evaluated for listening skills almost as much as speaking skills.

Answer Be genuine, without bragging. I've found in the past that students walk in like they own the world and think they know everything because they've had A+ in school. It's usually a very humbling experience when you start work and realize that you know next to nothing.

Answer Keep the university grades above average. Oftentimes, the only way to tell candidates apart (prior to an interview) is to look at the grades. Try and get some experience working for an insurance company while in university. Write a few exams to show your dedication and ability to write them.

Answer Have your résumé done professionally. Make sure there are no mistakes in it. Be prepared to relocate.

Answer Pass as many SOA exams as possible during the university year. It is easier to study at school than at work even if the employer gives us studying days. Be confident. Show interest. Do research on the company fields of practice, goals, successes, etc.

◆◆◆

3.2 | Moving Up the Ladder

Q How fast can an actuary expect to climb the ladder in your company? Illustrate your answer and compare it with examples from other companies you may be familiar with.

Answer There is not a typical path but I know that by the nature of the work, people in Asset Management become consultants and responsible for clients after fewer years than in the Retirement or Health and Welfare line of business for example. I can't compare with other companies.

Answer It depends on the number of exams and how devoted the actuary is.

Answer You start as an analyst and become a consultant after 6 to 10 years. Then you can become a senior consultant after another 5 years.

Answer In insurance companies, in order to climb the ladder, there must be "pace" up in the ladder. That is what I have observed, therefore the pace will be different for everyone. In consulting companies, such as Mercer and Hewitt, I have observed a more "parallel" way of going up. If one is good at what one does, then the possibility of moving up is there. Everyone seems to be given the same chance and I believe that climbing the ladder in the consulting business can be faster than in an insurance company where the years of dedication and work are usually rewarded.

Answer It's all based on exam performance. It is possible to pass exams too quickly, i.e., get your Fellowship without much work experience (under 5 years). A company would probably be reluctant to promote someone to management in that case. But it's still never a bad thing to pass exams, even if your work experience isn't commensurate with your exam success. That's a better spot to be in than to have a lot of experience, but struggling with exams.

Answer Will generally depend on quality of work, understanding issues, learning quickly, time and effort, ability to solve problems (find solutions), good communication skills.

Answer Depends on the number of hours willing to put in, rapidity in passing exams, and the efficacy of your work.

Answer Being an actuary is irrelevant to any ladder climbing I could do here. On the contrary, I should become more and more of a HR generalist if I wanted to go much higher.

Answer From new hire to Assistant Vice-President (officer) in ten years. Not at all uncommon in six or seven, depending on exam progress.

Answer In Canada, for P/C companies, most actuaries are at least managers by the time they are 30, with a fair number even being VP. Again, it all depends on the individuals, and their willingness to take risks. One may have to move to where the opportunities are, for example. It also depends on the size of the company. Opportunities at my prior company were limited, since there was a fair number of Fellows employed there. Still, with my departure, one opportunity was created. Sometimes, it is also a matter of being at the right place, at the right time.

♦♦♦

3.3 | Salaries and Benefits

Q How do you negotiate your salary, and have you always been satisfied with the salary and benefit packages you have had? Illustrate your answer from your knowledge of the practice of specific companies.

Answer I never really negotiated salaries and things like that since Towers Perrin has their own way to give salary increases and so on. So far I have felt satisfied with the way I was treated "money-wise," but I would not be afraid to discuss it with my supervisor if that was the case.

Answer Companies offer packages they believe are fair, and apply internal equity. Rarely are they negotiated.

Answer It is important to be aware of the market when negotiating. I used the D. W. Simpson website. It's a recruiting company that specializes in actuaries. They often run surveys on salary by exams and years of experience and post their results on their site. Also, it is important to know the company you're applying for. The size of the company and the field will influence the salary (insurance versus consultation, small company from Montreal versus international company). It is important to remember that not only the salary counts. The other benefits (health plan, stock purchase program, bonuses, study days), responsibilities, chances of promotions and environment, for example, should all be considered. So far, I've always been satisfied with my salary and benefits.

Answer Not negotiated. I accepted an offer and did not really test the waters with other companies.

Answer It is hard to negotiate salary and I don't think my employer offers a competitive package. They actually tell us that we are paid on "average" and not on the high salaries. Bonuses are great though (from 15% at hire to 25% for consultants).

Answer   As an intern in companies, I am not in a position to negotiate my salary, but I do know what to expect. Depending on the number of exams passed, the number of years in school, and the work experience, one's salary can vary. For my part, I have been very satisfied with the salary I have received.

Answer   I generally don't negotiate mine unless I'm changing companies. Then I negotiate to the point to which the salaries are roughly consistent across offers and I am picking my job based on the job, rather than the compensation. I used to work in the United States and they pay more. But they sometimes low-ball Canadian students who don't know the U.S. actuarial student market. I took a pay cut to return to Canada. So salary isn't the absolutely highest priority I have. I think that, given the lower cost of living in Canada, the standard of living of American and Canadian actuaries is about the same. I think that insurance companies always offer a very competitive benefits package—medical, retirement, cafeteria, fitness center benefits—that many other companies just don't have. Larger companies can afford to have more comprehensive benefits than smaller ones, I think.

Answer   No real negotiation. Performance objectives are set in advance and salary is determined based on reaching those objectives. I have been satisfied with my salary most of the time.

Answer   Tough to do because there is no one to compare to. Firms that hire many actuaries (consultants, insurers) have a good knowledge of the market, but others don't. Not that I have always been satisfied (who is?), but in this field, money isn't everything. You have to look at the quality of life also.

Answer   A new student has little flexibility. A student hired from another company has more ability to set the salary. Usually the hiring company will want to pay no more than 10% or 20% above what you are currently getting (this will depend on how long you have been in your current job and how hot the company is to get you). Remember that although you are an attractive commodity, if you overprice yourself you had better be able to deliver superb value or risk a reputation as a bad deal.

Answer   I start with the assumption that the company that I work for treats me equitably. If I discover that this is not the case, I don't hesitate to make a move. Until now I believe that only one of my employers (I have had three so far) was not paying me what I was worth, and I left him after nine months. Even though he wanted to correct the situation (with a 30% increase), I quit because my relationship with my employer is one of mutual trust.

Answer At the entry level and intermediate positions in my company, salaries are quite competitive with the market. Negotiation usually happens at the time of hire and future increases are based on merit, responsibility, level, and performance. Since I work for a very large employer with many people at or around the same level, I believe salaries are fair and competitive. Unless you are a superstar, there are probably only minor salary negotiations. Of course if you are good and the competition actively recruits you, you will most likely receive higher pay as an incentive.

Answer In my experience, I never had to negotiate much for my salary. Headhunters are usually a great help in negotiating on your behalf with a new company. I must admit I have also generally been satisfied with my salary and benefits, although other people out there may have better conditions. In turn, I'm probably doing better than other people out there. At the end of the day, what really counts is being happy with what you do and whom you work with.

Answer For positions that are non-managerial, companies usually have salary scales, which they usually follow. There is not a lot of room for negotiation. For managerial positions, salary is usually based on the experience and the skills of the candidate, as well as "current salary" as a basic input. This is done on a case-by-case basis, there is no general rule. At the executive level, a company will usually want a specific individual. Therefore, there is much more room for negotiation. There obviously is no general rule applicable in this case. Benefit packages are always very good.

Many employers have actuarial training programs for actuarial students. A typical training program will include assistance with exam preparation, rotational assignments to expose to the student to different aspects of actuarial work, seminars, purchase of study material, paid study time, and reimbursement of the costs associated with the writing of the examinations. The survey illustrates the value of such training programs for career development.

◆◆◆

Q Based on your employment experience, describe the support different companies give to students to prepare for actuarial examinations (study days, payment of examination fees, purchase of study material, etc.)

Answer Most employers who retain the services of actuaries generally provide full support.

Answer We have great support from the firm: three study days per hour of exams, all fees/books/study guides are paid for, we can ask for seminar support if needed, and we have a passing bonus in dollars and a pay increase too that are very good when compared to the market. Lots of support from fellow workers too. So our workload is not as huge in the final weeks of studying.

Answer   Most provide the same level of study days, three days per hour of exams, payment of fees, and study material. Very few will pay for seminars.

Answer   Usually the payment of examination fees, the study seminar support, and the purchase of study material and study aids are mostly the same from company to company. What I find the most important is the amount of study days (or the respect of your study time). This is what varies the most. Sometimes, you have to work overtime and on weekends to be able to take your study day... and often you can't even take your study day because you have a rush. This applies in the consultation field, it's not as true in the insurance field and doesn't apply at all if you work for the government. Instead, they should set a maximum number of working hours when you're studying.

Answer   For interns, most companies offer the same studying benefits. We usually get six study days, and the day of the exam off, and while some companies pay for books and the exams, others will only pay for the exam if you have successfully passed it.

Answer   Study days are generally based on the number of hours for the exam (that is, three days per hour up to 18 or 20 per session). Usually the exam fees are covered for the first (sometimes second) writing of an exam. After that they may be reimbursed upon passing of the exam only. The study materials and aids are generally purchased by the company.

Answer   We have a good study-time program. About 15 to 18 days for exams 5 [2002] and over. They pay 100% of the material and exam fees unless you fail too many times (I don't know how many!). They don't pay for support seminars.

Answer   I have found that most insurance companies are pretty similar in their student program. All pay the exam fees, study materials, and give a certain amount of studying days off to the student. If the student fails a course on his second attempt, the studying time is shorter but the other advantages stay the same. Most companies will stop paying exam fees, study time, and material after the third trial but will reimburse the fees if the exam is passed. Companies also offer bonuses upon successful completion of a course. I have found this to be very much the same in all companies, both in insurance and consulting companies.

Answer   Exam support seems to be pretty universal across the different companies. All companies offer study time, payment of exam fees, materials, etc. Some might pay for exam seminars. The study time probably won't vary from company to company too much, although whether you *actually* get to use all of your time is different. Insurance companies stress passing exams more so their students usually get their full study time allotment. Consulting companies usually work their students to the point that they lose out on some of their time or they have to try to cram it all at the end. If a company offers more study time, then they usually have a higher standard of required successful exam attempts to stay in their program.

Answer Generally: three study days per hour of examination paid by the company, examination fees paid by the company, study material and study aids paid by the company, study seminar may require approval. For unsuccessful candidates: for a third try no study days are provided (must take vacation days), only half of examination fees are paid and the other half is reimbursed if the exam is passed.

Answer The information below is based on my personal experience and what I have heard from actuaries in other companies. Most companies tend to give approximately 15 study days for a candidate's first attempt at a given exam. The variation between companies tends to come more for the second attempt, where I have seen anywhere from 7.5 to 12 days. Third and fourth attempts can be from 0 to 7.5 days. Most companies also tend to pay for the exam fees at least once and usually twice, if the candidate passes on the second attempt. Third or fourth attempts are often paid by the candidates themselves. Study materials are usually provided by the employer.

Answer Study days are a must, and you may have to negotiate for them if you join a small company. Sometimes fees are always paid no questions asked; sometimes they are paid upon a passing grade; sometimes they are paid for the first two tries, then 50% on the third try, then you're on your own, my friend. I've seen a wide variety depending on the companies.

Answer Consulting firms tend to give more support than insurance companies. Mercer seems to be the best of them all. They give paid study days and pay for all the necessary exam materials.

Answer Fifteen days study time, all fees paid for first attempt, partial for repeat, study aids paid for, company pays for seminar costs, travel for PD [professional development], etc.

Answer With respect to my company: study days are three days per hour of exam, cost of exam paid up front, study materials are paid up front. Students can keep their study aids (ACTEX, JAM, etc.), but the books must be returned to the study library; study seminars, up to 50% of the cost is paid up front, and the remainder is paid upon passing the exam.

Answer Most companies will give about fifteen paid study days per four hours of an exam, pay for successful completion of an exam, and buy the study material. Some companies will pay for part or all of a study seminar. Most companies will now penalize employees who consistently fail exams by reducing the number of study days provided upon the second and third failure of the same exam.

Answer Twelve to 16 study days. Exam fees, material, and study aids reimbursed/paid up front for first two attempts. Salary increase upon successful completion of an examination. Salary increase when the ACAS and FCAS designations are attained. Partial reimbursement for out-of-town study seminars. Cash bonus if successful on first attempt.

◆◆◆

3.4 | Company Reputation

In this section, several respondents of the survey comment on the quality of the work environment of past and current employers.

Q **Are there companies with a good or bad reputation in the actuarial industry? Give some examples and explain why.**

Answer Mercer Human Resource Consulting has a tremendously good reputation in the industry as a fair and rigorous training ground for younger professionals. Furthermore, their global presence provides valuable opportunities for transfers.

Answer I have found that at Towers Perrin people really enjoy being in the environment we have here. Lots of flexibility from everyone, working hours and holidays, senior people help the younger ones a lot, and I really appreciate the support I am getting. Very few people have left the company since I have been here. I can't really speak about other companies.

Answer Consulting actuaries are known to work a lot, maybe too much. They depend on clients and want to do everything by yesterday because they think the clients will be more satisfied. But it's a challenging job. You get to work with clients and in teams, where interpersonal skills are very important. At the other end of the spectrum, actuaries working for the government are just like other white collar workers. They may not always work a lot, may not have high salaries, and may do boring or repetitive jobs. But they will have more time to study and so will probably become Fellows faster.

Answer Consulting firms are recognized to make their employees work a lot but from my personal experience, the workload gets better after a few years.

Answer Most major companies are respected by the others. I think the major companies realize that all companies need to have a good reputation so that the industry itself does not suffer.

Insurance companies sell intangible goods—they sell promises. They promise to pay money when they're contractually obligated to. So if people start believing that insurance companies can't be trusted to keep their promises, then every company suffers.

Companies may have differences of opinion on what qualifies as ethical sales methods—policy replacements and churning. I'm reluctant to name any specific company with a bad reputation because my experience is that unethical behavior is often restricted to isolated cases of miscommunications or overzealous field agents, rather than an unethical systemic problem knowingly practiced by the company.

Answer Yes and no. Each company has its own set of core values that are disseminated by more senior employees. These values will fit with some people, clients, etc. Hence, there will always be a need for different companies to fit everyone's needs.

Answer Reputations change over time. Big companies tend to attract good students and spend a lot of time training them. They cannot risk bad publicity. Small companies are often reputed to underbid.

Answer Consultants had a bad reputation of making you work very hard, but things have changed and now they are fighting over the new graduates. What's tough about consulting is that you are told what to do (not much room for your own input before you reach the strategic consultant level). I don't find that I work less since leaving the consulting business. It's just that now, my work is more focused and I have much more influence over the decision-makers.

Answer Mercer is outstanding. They hire only the best people so it creates a great work/learning environment. Who better to learn from than the best in their field?

Answer Manulife has a good reputation because we have a lot of actuaries and we are a big company with many opportunities.

Answer Consulting firms usually have the reputation of being "sweatshops." To a certain extent, consultants may put in more hours than insurance actuaries, for example, but the pay at higher levels is also more significant. Consulting actuaries work more than nine-to-five, but the hours can be flexible, and calling them sweatshops is probably an exaggeration.

Answer   I'm not sure that there are companies with a good or bad reputation in the same way in which there are individuals with good or bad reputation. Not all actuaries are qualified to be managers. However, by virtue of their Fellowship, most actuaries do become managers. Whereas some of them are just not good managers, some others are actually bad managers. By this I mean that people will work for them for short periods (usually less than a year) and move on to another company. This results in the company having a revolving door (and a bad reputation), where actuaries leave every few months.

In order to find a company with good managers, students should to talk to other students, ideally ones who have just left a particular company, or ones who are currently working there, before deciding to change jobs. If you find yourself, each and every morning, not wanting to go to work, you probably have a bad manager, and I would advise changing companies. It may also be that an actuarial career is not for you, in which case you should change careers.

◆◆◆

3.5 Consulting versus Insurance

Actuarial careers can be looked at in several ways. Consulting versus insurance is one of them. Here are some comments from respondents to the survey about the advantages of these career options:

Q **What are the main advantages of working for (a) a consulting firm, (b) an insurance company?**

Answer   Consulting firms: flexible schedule, diversity of work, relations with clients (sometimes a downside), lots of learning opportunities and career paths, meeting interesting people with different expertise and backgrounds.

Answer   Consulting offers better long-term compensation conditions, and is more client-oriented.

Answer   In an insurance company you will generally have a shorter day of work. The work you have to do is for your company so you will be less squeezed with urgent deadlines. For me, the main advantage is that you will be working with people who have the same skills as you. If you have something to explain, it will be to someone that understands how actuarial mathematics works. The main advantage of a consulting firm is the contact with clients. It makes the work more different from one day to another. Also, you might be asked to develop new programs, thus you are constantly learning new stuff.

Answer Insurance companies: more stable work schedule, salary increase for each exam pass, more technical work at junior levels. Consulting firms: more communication involved such as meeting with clients, challenged by peers about current issues, etc.

Answer Insurance companies: job stability. The regular hours and more relaxed atmosphere. Working in an insurance company requires more technical skills so if someone prefers the technical work, that's their place! Consulting firms: the interaction with clients, the variety of the work, and the chance to work in larger teams.

Answer Having worked in both, I believe that it boils down to who you are working for, the team, and what work is available. The work is often very similar. Insurance companies: opportunity to do more types of projects than just loss reserves and rate filings. For those who will pass exams more slowly, probably better opportunities. Consulting firms: tend to attract and retain those who are bright and willing to work a bit harder for the money (although those at insurance companies can work as hard or harder). Can be more variety among types of projects (reserves, rates, types of clients). Generally higher standards, especially for documentation. Generally faster promotions for those who pass exams. Fewer administrative meetings. Projects tend to have a definite end (unlike insurance companies, where projects can be put on the back burner forever).

Answer Insurance companies: more exam support, pass through exams quicker, get promoted more quickly as a result, rotation program in different areas of the company. Consulting firms: higher initial salary and ultimately more rewarding if you can manage to pass your exams quickly.

Answer Consulting firm advantages: higher salary, exposure to a wider range of actuarial topics. Greater sense of ownership for your work as it affects the results of your company. Consulting firm disadvantages: longer hours, more pressure, sometimes difficult to fit in study days.

Answer Insurance companies: stable workflow, job you keep for a long time. Consulting firms: stimulating work and younger colleagues!

Answer In a consulting firm you work longer hours, but you have the advantage of being your own boss. You work the hours that you would like to. In a consulting firm there is less of a hierarchical structure. Everyone who merits it will get to be a boss/consultant. There is more of an entrepreneurial spirit since you work directly with your clients and there is more of a personal profit incentive.

Answer As a student, you will touch more varied projects working for a consulting firm. There is also a lot less programming and data entry involved with consulting firms. However, the best consultants are the ones who understand how an insurance company works, and what are the challenges of running a company. The only way to learn this is to work for an insurance company. Once higher up in the organization, being involved with the day-to-day management of a company is also a rewarding experience, which consulting cannot offer.

Answer   Insurance companies: lower workload (apparently), less pressure to produce (apparently), better knowledge of the business because of focus on one company/group of companies, more involvement in decision-making process, comfort derived from knowing the environment and the people. Consulting firms: higher income, higher standards, no interdepartmental politics.

Answer   Insurance companies: less stress, more flexible schedule, fewer work hours, social benefits more generous. Consulting firms: more challenging, possibility to have a promotion, higher salary, more diversified work.

CONSULTING FIRMS

In various reviews published regularly on the Internet, by Kennedy Information LLC, for example, human resource consulting firms are often ranked according to their levels of revenue. According to one recent ranking, compiled by Workforce Management in 2007, Mercer Human Resource Consulting (1), Deloitte Consulting (2), Watson Wyatt Worldwide (3), Aon Consulting (4), PriceWaterhouseCoopers (5), Towers Perrin (6), and Hewitt Associates (7) were listed as the top seven firms. Since 2007, the industry has gone through a number of consolidations with the mergers of Towers Perrin and Watson Wyatt to form Towers Watson and Aon with Hewitt to form Aon Hewitt. All of these firms are included in the profile below, albeit from a career perspective.

The rationale for the inclusion of the remaining firms in this guide is based on personal familiarity with these companies acquired as the long-time director of the actuarial cooperative education program at my university. At the beginning of their careers, actuarial students and actuaries working for these companies are small fish in big ponds. As their careers progress, therefore, some choose to advance their careers by becoming big fish in smaller ponds. Once trained by larger companies, many human resources consultants tend to break off from their larger employers to form their own practices. The companies listed, therefore, also profile a number of smaller companies.

Most company websites now include material that helps candidates narrow down their career choices: the mission statements, published career types and options, and other resource material now make it easy to develop a sense of relevance. Some of this information is included in the company profiles below if it adds to a generic sense of what actuaries do in the real world. For completeness, the listed company websites should be consulted in all cases. Useful information about all of the companies below can also be found on the Facebook, LinkedIn, and Twitter pages of these companies.

A.1 | Aon Hewitt Corporation

- **Company Website:** www.aonhewitt.com/
- **Careers:** www.aonhewittcareers.com
- **Facebook:** www.facebook.com/AonHewitt
- **LinkedIn:** www.linkedin.com/company/aon-hewitt
- **Twitter:** twitter.com/#!/AonHewittCareer

Aon is a Fortune 500 company that is a world leader in risk management, retail, reinsurance and wholesale brokerage, claims management, specialty services, and human capital consulting services. The company has an employee base of 55,000 people working in 600 offices in more than 125 countries. In the United States, Aon has offices in all states. Aon Corporation is headquartered in Chicago and is a leading provider of risk management services, insurance and reinsurance brokerage, and human capital and management consulting.

Careers

Unite with Aon: Every great team depends on the individual contributions of its members; the skills, talents and the passion that each and every team member brings to every game. At Aon, thats how we achieve our goals. We have a passion for excellence because we know our work touches the lives of millions of individuals all around the world every day. Employees. Families. Companies. Retirees. As the Principal Sponsor of Manchester United—the most admired sports brand on the planet—Aon proudly partners with a world-class team. What do Aon and Manchester United have in common? Manchester United shares Aon's passion for excellence and our desire to win—based on teamwork, tradition, integrity, competition, and success. Like Manchester United, Aon is passionate about achieving goals. Were especially passionate about helping our clients—and our employees— reach theirs. And, like Manchester United, were making our mark on the world, one goal at a time.

Locations

Africa: Angola, Botswana, Ghana, Kenya, Lesotho, Malawi, Morocco, Mozambique, South Africa, Swaziland, Tanzania, Tunisia, Uganda, Zambia, and Zimbabwe. **Asia–Pacific:** Australia, China, Fiji, Guam, Hong Kong, India, Indonesia, Japan, Korea, Malaysia, New Zealand, Papua New Guinea, Philippines, Saipan, Singapore, Taiwan, Thailand, Vanuatu, and Vietnam. **Europe:**

Austria, Azerbaijan, Belgium, Bulgaria, Croatia, Cyprus, Czech Republic, Denmark, Estonia, Finland, France, Georgia, Germany, Gibraltar, Greece, Guernsey, Hungary, Ireland, Isle of Man, Italy, Kazakhstan, Latvia, Lithuania, Luxembourg, Malta, Netherlands, Norway, Poland, Serbia, Slovak Republic, Slovenia, Switzerland, Turkey, and the United Kingdom. **Middle East:** Bahrain, Egypt, Israel, Jordan, Lebanon, Oman, Pakistan, Qatar, Saudi Arabia, United Arab Emirates, and Yemen. **North America:** Antigua, Aruba, Bahamas, Barabados, Belize, Bermuda, British V.I., Canada, Cayman Islands, Costa Rica, Dominican Republic, El Salvador, Grenada, Guatemala, Haiti, Honduras, Jamaica, Mexico, Netherlands Antilles, Panama, Nicaragua, Puerto Rico, Trinidad & Tobago, Turks & Caicos, and the United States; **South America:** Argentina, Bolivia, Brazil, Chile, Colombia, Ecuador, Guyana, Peru, Suriname, Uruguay, and Venezuela.

A.2 | BNY MELLON

- **Company Website:** www.bnymellon.com/
- **Careers:** www.bnymellon.com/careers/
- **Facebook:** www.facebook.com/bnymelloncareers
- **LinkedIn:** www.linkedin.com/company/bny-mellon
- **Twitter:** https://twitter.com/BNYMellon

BNY MELLON is headquartered in New York City. It is a multinational banking and financial services company. The company was established in 2007 from the merger of Mellon Financial Corporation and The Bank of New York Company, Inc. It is a leading investment management and investment services company and has $25.8 trillion in assets under custody or administration and $1.26 trillion under management. It lists Foresight, Asset Management, Security Servies, Treasury Services, and Wealth Management as its main areas of activities. The company's clients include local companies and global corporations, not-for-profit and educational institutions, and numerous state and local governments. With offices in 36 countries, the company provides *financial services for institutions, corporations and high-net-worth individuals in more than 100 markets around the world.*

Careers

BNY MELLON describes itself as a company that is *a performance-driven, global organization underpinned by a spirit of teamwork and trust*; has *a focus on inclusion and results, and a dedication to the creation of innovative financial services and outperformance*; and provides *leadership, support and growth opportunities to help you realize your full potential.*

The company reassures its potential employees that they will *have access to myriad career paths in a variety of disciplines. From asset management and wealth management to securities servicing and treasury services. And, in a host of corporate support disciplines, such as finance, technology, human resources, risk, compliance, legal, marketing, communications and more.*

Locations

In addition to its multiple offices in the United States, the HRIS division of Mellon has international offices in Adelaide, Barcelona, Brisbane, Bristol, Brussels, Dublin, Edinburgh, Gouda, Hong Kong, Houston, Ipswich, London, Madrid, Manchester, Melbourne, Mexico City, Montreal, Ottawa, Paris, Perth, Reading, Singapore, Sydney, Toronto, Vienna, Warsaw, and Wiesbaden.

A.3   Buck Consultants

- **Company Website:** www.buckconsultants.com
- **Careers:** www.simplyhired.com/a/jobs/list/c-buck+consultants
- **Facebook:**
  www.facebook.com/pages/Buck-Consultants/109619939063524
- **LinkedIn:** www.linkedin.com/company/buck-consultants
- **Twitter:** twitter.com/BuckConsultants

Buck Consultants is a Xerox company offering integrated services in communication, compensation, global investment consulting, global technology and delivery solutions, health and productivity, talent and HR solutions, and retirement. Buck Consultants is credited with a number of firsts, including the establishment of the first global employee stock ownership plan and the development of the first fully integrated health savings account (HSA).

Careers

Buck Consultants, like some other consulting firms, runs its own educational program at Buck Consultants University (BCU) that *provides Buck staff with cross-functional development opportunities for a broad-based consulting career. The syllabus is composed of technically focused sessions designed to span the career of a consulting professional—from foundational skills for interns and entry-level staff to specific topics for the experienced consultant. Our offerings include hundreds of technical and soft skills courses that will help you develop a customized learning plan to meet your development needs.* The offered courses are divided into "colleges" specializing in areas such as retirement, health and productivity, compensation, communication, human capital management, and system education. The *Actuarial Student Academic Program* provides actuarial students with *time off with pay to prepare for exams, financial support for study materials and seminars, and rewards for successful completion of exams and designations.*

Locations

In the United States, Buck Consultants has offices in Atlanta, Boston, Chicago, Cincinnati, Cleveland, Dallas, Denver, Detroit, Fort Wayne, Honolulu, Houston, Los Angeles, Maumee, Melville, Minneapolis, New York, Orange County, Philadelphia, Phoenix, Pittsburgh, San Diego, San Francisco, Secaucus, St. Louis, Stamford, Tampa, and Washington, D.C. In addition, Buck Consultants has international offices around the world including Latin America, Europe, Asia-Pacific, and Africa.

A.4 Deloitte

- **Company Website:** www.deloitte.com/
- **Careers:** www.deloitte.com/Careers/
- **Facebook:** www.facebook.com/deloitte
- **LinkedIn:** www.linkedin.com/company/deloitte
- **Twitter:** twitter.com/deloitte

Deloitte is headquartered in New York City. According to its own website, *in the United States, Deloitte LLP and its subsidiaries have 45,000 professionals with a single focus: serving our clients and helping them solve their toughest problems. We work in four key business areas—audit, financial advisory, tax and consulting—but our real strength comes from combining the talents of those groups to address clients, needs. Fortune and BusinessWeek consistently rank our organization among the best places to work, which is good news for our talent and our clients alike. When the best people tackle the most compelling challenges, everyone wins.*

Deloitte is the brand under which tens of thousands of dedicated professionals in independent firms throughout the world collaborate to provide audit, consulting, financial advisory, risk management and tax services to selected clients. These firms are members of Deloitte Touche Tohmatsu Limited (DTTL), a UK private company limited by guarantee. Each member firm provides services in a particular geographic area and is subject to the laws and professional regulations of the particular country or countries in which it operates. DTTL does not itself provide services to clients. DTTL and each DTTL member firm are separate and distinct legal entities, which cannot obligate each other. DTTL and each DTTL member firm are liable only for their own acts or omissions and not those of each other. Each DTTL member firm is structured differently in accordance with national laws, regulations, customary practice, and other factors, and may secure the provision of professional services in its territory through subsidiaries, affiliates and/or other entities.

Careers

Deloitte puts it this way: *What's standing between you and your dreams? Just one line on your resume. Our reputation for providing high quality services with integrity has earned us the trust of our clients, and our people. If you're ready for a career with a dynamic organization in an environment that fosters professional development and career advancement, you're ready for Deloitte. With 182,000 people in over 150 countries, Deloitte member firms serve more than 80 percent of the world's largest companies as well as large national enterprises, public institutions and successful fast-growing companies.*

Locations

In the United States, Deloitte has over 100 locations. Internationally, it is represented by divisions of Deloitte Touche Tohmatsu Ltd. in Albania, Angola, Argentina, Aruba, Australia, Austria, Azerbaijan, Bahamas, Bahrain, Barbados, Belarus, Belgium, Bermuda, Bosnia and Herzegovina, Brazil, British Virgin Islands, Brunei, Bulgaria, Canada, Cayman Islands, Chile, China, Colombia, Costa Rica, Croatia, Cyprus, Czech Republic, Denmark, Dominican Republic, Ecuador, Egypt, Estonia, El Salvador, Finland, France, Gaza Strip/West Bank, Georgia, Germany, Gibraltar, Greece, Guam, Guatemala, Honduras, Hungary, Iceland, India, Indonesia, Ireland, Israel, Italy, Jamaica, Japan, Jordan, Kazakhstan, Kenya, Korea, Kuwait, Kyrgyzstan, Latvia, Lebanon, Libya, Lithuania, Luxemburg, Malaysia, Malta, Mexico, Moldova, Montenegro, Morocco, Mozambique, Namibia, Netherlands, Netherlands Antilles, New Zealand, Nicaragua, Nigeria, Norway, Oman, Pakistan, Panama, Papua New Guinea, Paraguay, Peru, Philippines, Poland, Portugal, Qatar, Romania, Russia, Saudi Arabia, Serbia, Singapore, Slovak Republic, Slovenia, South Africa, Spain, Sweden, Switzerland, Syria, Taiwan, Thailand, Tunisia, Turkey, Ukraine, United Arab Emirates, United Kingdom, Uruguay, Uzbekistan, Venezuela, Vietnam, and Yemen.

The Canadian division of Deloitte, Deloitte & Touche LLP, is the Canadian member firm of Deloitte Touche Tohmatsu Ltd., which is a network of member firms, each of which is a legally separate and independent entity. In Quebec, Deloitte operates as Samson Blair/Deloitte & Touche. It is *one of Canada's leading professional services firms, [and] provides audit, tax, consulting and financial advisory services to a wide range of Canadian and international clients.*

The Canadian locations include Alma, Amos, Brossard, Burlington, Bécancour, Calgary, Chicoutimi, Dolbeau-Mistassini, Drummondville, Edmonton, Farnham, Granby, Grand-Mere, Halifax, Hawkesbury, Jonquière, Kanata, La Sarre, Langley, Laval, London, Magog, Mantane, Mississauga, Montréal, Nicolet, North York, Oakville, Ottawa, Prince Albert, Prince George, Québec City, Regina, Rimouski, Roberval, Rouyn-Noranda, Saint John, Saint-Hyacinthe, Saskatoon, Sept-Îles, Sherbrooke, Sorel, St-Félicient, St. Catharines, St. Johns, Toronto,

Trois-Pistoles, Trois-Rivièeres, Val D'Or, Vancouver, Victoria, Ville de la Baie, Windsor, and Winnipeg.

A.5 Dion Durrell

- **Company Website:** www.dion-durrell.com
- **Careers:** www.dion-durrell.com/en/careers
- **Facebook:** www.facebook.com/pages/Dion-Durrell
- **LinkedIn:**
  www.linkedin.com/company/dion-durrell-associates-inc

Dion Durrell is headquartered in Toronto, Canada. The company is an actuarial and insurance consulting firm. It creates and implements *innovative strategies that encompass risk financing, insurance management and insurance distribution solutions.*

Careers

Dion Durrell sees itself as a company that has a strong desire to see its associates fulfilled by their work and their work environment. The company states that *we believe you are at your best when your contribution is valued and when you can use your unique talents, every day. Since our inception in 1995, we have created a culture that makes this possible—as it fosters the career development of every Dion Durrell Associate.*

Strongly entrepreneurial, we value individuals who can think independently as they work together with like-minded Associates to create innovative, unique and value-added solutions. We pride ourselves on an atmosphere that cultivates mutual commitment between Associates, as well as between the firm and the many blue-chip clients whose challenging problems we are asked to solve. We reward your efforts with competitive compensation and benefits, and invest in your future through continual career and professional development.

Locations

Dion Durrell has offices in London (UK), Montreal, Oakbrook Terrace (Illinois), St. Michael (Barbados), and Toronto (Canada).

A.6 Ernst & Young

- **Company Website:** www.ey.com
- **Careers:** www.ey.com/Careers
- **Facebook:** www.facebook.com/pages/Ernst-Young

- **LinkedIn:** www.linkedin.com/company/ernst-and-young
- **Twitter:** twitter.com/ernst_and_young

Ernst & Young is headquartered in New York City. As stated on its website, it offers *a broad array of solutions in audit, tax, corporate finance, transactions, online security, enterprise risk management, the valuation of intangibles, and other critical business-performance issues.*

Careers

Over three months in 2007, Ernst & Young built the fastest-growing recruitment group on Facebook, and it now has more than 12,200 members. The company has over 25,000 employees in the U.S. and over 88,000 elsewhere. in 2008, CCMoney ranked it among the best 100 companies to work for. For students, the company offers internships and full-time positions with rewarding and challenging opportunities that can serve as launchpads for promising careers. For experienced actuaries, it offers *a dynamic environment* that will help candidates achieve their potential. For executives, finally, Ernst & Young suggests that *if you are a proven leader in your field, with an entrepreneurial mindset and a passion for helping your clients achieve their potential, you can help us shape the future of our business.*

Locations

In the United States, Ernst & Young has over 70 offices. You will find them in Atlanta, Austin, Baltimore, Birmingham, Boca Raton, Boston, Buffalo, Charleston, Charlotte, Chattanooga, Chicago, Cincinnati/Dayton, Cleveland/Akron, Columbus, Dallas, Denver, Des Moines, Detroit, Ft. Worth, Grand Rapids, Greenville, Hartford, Honolulu, Houston, Indianapolis, Iselin, Jacksonville, Jericho, Kansas City, Las Vegas, Los Angeles, Louisville, McLean, Meadowlands/ Secaucus, Memphis, Miami, Milwaukee, Minneapolis, Nashville, New Orleans, New York, Oklahoma City, Omaha, Orange County, Orlando, Palo Alto, Philadelphia, Phoenix, Pittsburgh, Portland, Providence, Raleigh, Richmond, Rochester, Rogers, Sacramento, Salt Lake City, San Antonio, San Diego, San Francisco, San Jose, San Juan, Puerto Rico, Seattle, St. Louis, Stamford, Syracuse, Tampa, Toledo, Tulsa, Washington, D.C., and Westlake Village. In addition, the company has offices in more than 130 countries.

A.7 | Hay Group

- **Company Website:** www.haygroup.com/ww/
- **Careers:** www.haygroup.com/ww/careers/
- **Facebook:** www.facebook.com/HayGroup

- **LinkedIn:** www.linkedin.com/jobs/c-Hay-Group
- **Twitter:** twitter.com/Hay_Group

The Hay Management Consulting Group is an independently owned company. The company states that it *is a global management consulting firm that works with leaders to transform strategy into reality. We develop talent, organize people to be more effective and motivate them to perform at their best. Our focus is on making change happen and helping people and organizations realize their potential.*

We have over 2,600 employees working in 84 offices in 48 countries. Our clients come from the private, public and not-for-profit sectors, across every major industry and represent diverse business challenges. For over 60 years, we have been renowned for the quality of our research and the intellectual rigor of our work.

Hay Group's actuarial services form the backbone of their retirement plans practice. Hay Group actuaries, who are qualified under the requirements of the Society of Actuaries and the federal Joint Board for the Enrollment of Actuaries, provide the entire range of services from annual actuarial valuations to valuations for mergers and spin-offs to plan termination valuations.

Careers

Hay Group encourages individuality and initiative and supports its people *through continuous development and feedback.* The company says that it wants its people *to have a life outside work and respect the job/life balance. For some this means flexible working arrangements and for others its about making the most of our flexible benefits offerings.*

Hay Group people can shape their careers in the manner that they choose. Career progression is based on performance, with clearly defined expectations and a performance management process to help shape individual growth. This is backed by training and skills development which draws on both internal and external opportunities and providers. Educational assistance is also available to those seeking to further qualifications.

Locations

North America: Canada, Costa Rica, Mexico, and the United States. **South America:** Argentina, Brazil, Chile, Colombia, Peru, and Venezuela. **Africa:** South Africa. **Europe and Middle East:** Austria, Belgium, Czech Republic, Finland, France, Germany, Greece, Hungary, Ireland, Israel, Italy, Lithuania, Netherlands, Norway, Poland, Portugal, Romania, Russia, Slovak Republic, Spain, Sweden, Switzerland, Turkey, Ukraine, United Arab Emirates, and the United Kingdom. **Asia:** China, India, Indonesia, Japan, Malaysia, Singapore, South Korea, and Thailand. **Pacific:** Australia and New Zealand. In Canada, the Hay Group has offices in Calgary, Edmonton, Halifax, Montreal, Ottawa, Regina, Toronto, and Vancouver.

A.8   KMPG

- **Company Website:** www.kpmg.com/ca/en/whatwedo/pages/default
  .aspx
- **Careers:** www.kpmg.com/global/en/joinus/pages/default.aspx
- **Facebook:** www.facebook.com/KPMGRecruitment
- **LinkedIn:** www.linkedin.com/company/kpmg
- **Twitter:** twitter.com/kpmg

KMPG is headquartered in the Netherlands. Its actuarial competency derives from the fact that its actuaries have access to KPMG's global resources and to the extensive experience *gained from advising listed and unlisted companies, multinational enterprises, government entities and leading not-for-profit organizations.*

The Australian wing of KMPG, Actuarial Advisory Services, employs actuaries specializing in risk management. KMPG states that its Australian services are *widely used in the insurance and finance industries, and that actuarial analysis is entering into many other areas as its power and flexibility become more widely appreciated. Increasingly, actuarial skills are being sought to support multi-disciplinary approaches to important business decisions. Actuaries are experienced in both measuring and managing risk; although they cannot prevent irrational behaviour, actuarial methods can help to mitigate its impact and reduce uncertainties.*

The professionals at KMPG work *in general insurance, life insurance and wealth management, and health financing, although their services are being called on by an increasingly broad range of entities.*

KMPG actuarial services include *capital and risk management, planning and strategy, operations improvement, financial and statistical modelling, appointed/approved actuary, self insurance and risk management, litigation and arbitration support, internal control reviews, financial reporting, reserve review and analysis, pricing and product development, and merger and acquisition support.*

The company continues to say that its assignments *often involve long-term financial arrangements where risks and uncertain outcomes must be managed by the organization* and require *business modelling, valuation of obligations and product pricing, risk analysis and statistical analysis and interpretation.* KMPG points out that its recently established *Russia and the CIS Actuarial Services practice brings a new dimension to financial reporting and management of actuarial processes and future events in your insurance, employee benefits or pension fund arrangements in order to manage risk and improve cost.*

The *Russia and the CIS Actuarial Services practice is an extension of KPMG's Global Actuarial Services, underpinned by an international footprint of diverse actuarial practices, covering over 30 countries with over 300 professionals.*

*Members of our team are affiliated with local and international actuarial govern-
ing bodies such as the Institute of Actuaries in the United Kingdom, the International Actuarial Association and the Guild of Actuaries of the Russian Federation.
Our working relationship with KPMG's worldwide actuarial network combined
with a deeper knowledge of the local market enables us to deliver unparalleled
best practice actuarial services ahead of localized developments.*

Careers

The company points out that when you join KPMG, *you are not only joining
a network of more than 140,000 professionals in over 146 countries, you are
part of a team of people who are dedicated to client service excellence. Lever-
aging the skills, knowledge, financial resources, and passion of our firm and our
people allows us to serve our clients with uncompromising professionalism, cut-
ting though complexity to provide valuable insight in Canada and around the
world.*

Locations

KMPG has offices in over 150 countries. They include Afghanistan, Albania,
Algeria, Andorra, Angola, Anguilla, Antigua and Barbuda, Argentina, Armenia,
Australia, Austria, Bahamas, Bahrain, Bangladesh, Barbados, Belarus, Bel-
gium, Bermuda, Bosnia and Herzegovina, Botswana, Brazil, British Virgin
Islands, Brunei, Bulgaria, Cambodia, Cameroun, Canada, Cayman Islands, Chan-
nel Islands, Chile, China, Colombia, Congo, Cook Islands, Costa Rica, Croa-
tia, Cyprus, Czech Republic, Denmark, Dominican Republic, Dutch Caribbean,
Ecuador, Egypt, El Salvador, Estonia, Fiji Islands, Finland, France, French
Antilles, French Guiana, French Polynesia, Georgia, Germany, Ghana, Gibraltar,
Greece, Guatemala, Honduras, Hungary, Iceland, India, Indonesia, Ireland, Isle
of Man, Israel, Italy, Ivory Coast, Jamaica, Japan, Jordan, Kazakhstan, Kenya,
Kosovo, Kuwait, Kyrgyzstan, Lao, Latvia, Lebanon, Liechtenstein, Lithua-
nia, Luxembourg, Macedonia, Malawi, Malaysia, Maldives, Malta, Mauritius,
Mexico, Moldova, Monaco, Montenegro, Morocco, Mozambique, Namibia,
Netherlands, New Caledonia, New Zealand, Nicaragua, Nigeria, Norway, Oman,
Pakistan, Panama, Papua New Guinea, Peru, Philippines, Poland, Portugal, Puerto
Rico, Qatar, Romania, Russia, Rwanda, Saudi Arabia, Senegal, Serbia, Sierra
Leone, Singapore, Slovakia, Slovenia, South Africa, South Korea, Spain, Sri
Lanka, St. Lucia, St. Vincent and the Grenadines, Swaziland, Sweden, Switzer-
land, Syria, Taiwan, Tanzania, Thailand, Togo, Trinidad and Tobago, Tunisia,
Turkey, Turks and Caicos Islands, Uganda, Ukraine, United Arab Emirates,
United Kingdom, United States of America, Uruguay, Venezuela, Vietnam,
Yemen, Zambia, and Zimbabwe.

A.9   Mercer

- **Company Website:** www.mercer.com/
- **Careers:** www.mercer.com/careers
- **Facebook:** www.facebook.com/pages/Mercer/104155236288536
- **LinkedIn:** www.linkedin.com/company/mercer
- **Twitter:** twitter.com/#!/MercerInsights

Mercer Human Resource Consulting is headquartered in New York City. It is a Marsh & McLennan company. It is the world's largest human resources consulting firm. In 2011 and for several years before that, Vault.com ranked Mercer as the number one human resource consultancy firm in the world. According to the ranking, *Mercer's major competitors include Towers Watson, Buck Consultants, Aon Hewitt, and Hay Group.*

Careers

Mercer states that it *is the world leader in helping organizations leverage the power of their people to achieve peak company performance. If you thrive on challenge, are passionate about ideas, love solving problems, and truly enjoy connecting with people, we encourage you to explore the hundreds of job opportunities available through Mercer. Our core strengths in consulting, outsourcing and investments place Mercer in a unique position to help our clients achieve the extraordinary and extraordinary results require extraordinary people.*

Mercer states that working for it means

- *Working with the world's leading companies and organizations to help them meet their most critical business challenges—Mercer counts 9 out of 10 Fortune 100 companies among its clients.*
- *Working with the best and brightest people, colleagues who are committed to exceeding the high standards set by our clients—and by ourselves.*
- *Working for a company that offers a vast array of learning and development opportunities.*
- *Working for a company that conducts business with the very highest standards of integrity.*

Locations

In addition to its 43 American locations in Albuquerque, Atlanta, Baltimore, Birmingham, Boston, Charlotte, Chicago, Cincinnati, Cleveland, Columbus, Dallas, Deerfield IL, Denver, Detroit, Houston, Indianapolis, Kansas City, Los Angeles, Louisville, Memphis, Milwaukee, Minneapolis, New York, Norwalk, Orange, Philadelphia, Phoenix, Pittsburgh, Portland, Princeton, Richmond, Rochester, Salt Lake City, San Francisco, San Jose, Seattle, St. Louis, Tampa, and Washington, D.C., Mercer has international offices in over 40 countries around the world: Argentina, Australia, Austria, Belgium, Brazil, Canada, Chile, China, Colombia, Czech Republic, Denmark, Finland, France, Germany, Hong Kong, Hungary, India, Indonesia, Ireland, Italy, Japan, Malaysia, Mexico, New Zealand, Norway, Philippines, Poland, Portugal, Singapore, South Korea, Spain, Sweden, Switzerland, Taiwan, Thailand, the Netherlands, Turkey, the United Kingdom, and Venezuela. The company has more than 15,000 employees.

A.10 │ Normandin Beaudry

- **Company Website:** www.normandin-beaudry.ca
- **Careers:** www.normandin-beaudry.ca/carrieres/index.en.html
- **LinkedIn:** www.linkedin.com/company/normandin-beaudry

Many cities have small, often specialized, actuarial consulting firms that meet specific needs of their communities. In Montreal, one of these jewels is Normandin Beaudry Actuaires Conseil Inc. It is a private company categorized under Actuarial Consultant and located in Montreal, QC, Canada. Current estimates show this company has an annual revenue of $7,545,600 and employs a staff of approximately 110. The company aims to be the benchmark of actuarial consulting firms for Quebec's enterprises. It is active in pension and savings plans, group benefits, property and casualty insurance, risk management, asset management consulting, and financial commitment valuation.

Careers

Normandin Beaudry offers stimulating career opportunities that allow its employees to grow, both professionally and personally, in a work environment that promotes, the company's values of excellence, collegiality, initiative and respect for individuals. Many of the actuaries working at Normandin Beaudry are graduates of the author's actuarial co-op program and have confirmed that the harmony that exists among the company's employees sets it apart from other employers in the actuarial consulting field. Its employees are big fish in a small pond and many of them have become accomplished actuaries by being small fish in big ponds.

The corporate values of Normandin Beaudry are well-defined, embraced, and shared throughout the company. Normandin Beaudry strives for excellence: *discipline, objectivity, quality and reliability combine to produce excellence—our distinguishing characteristic.* The company *welcomes constructive criticism and values the cooperation generated by the sharing of knowledge and opinions.* Normandin Beaudry rewards initiative. It believes *that the greatest achievements are the result of a myriad of different ideas considered, discussed and selected.* The company respects individuality: *at Normandin Beaudry, respect is a two-way street.* The company believes *in individual and collective choices, and it is these choices that form the basis of a solid team.* As consultants and contributors, Normandin Beaudry adopts the "Normandin Beaudry attitude," which is summarized in the six words that comprise its credo and that guide it: *transparency, ethics and objectivity; and discipline, innovation and expertise.*

Location

Montreal.

A.11 | PricewaterhouseCoopers

- **Company Website:** www.pwc.com/
- **Careers:** www.pwc.com/gx/en/careers
- **Facebook:** www.facebook.com/pwccanada
- **LinkedIn:** www.linkedin.com/company/pwc
- **Twitter:** twitter.com/pwc_us_careers

PricewaterhouseCoopers (PwC) is headquartered in New York City. The worldwide services of PwC are organized into five main categories: (1) Audit, Assurance and Business Advisory Services, (2) Business Process Outsourcing, (3) Corporate Finance and Recovery Services, (4) Human Resource Services, and (5) Global Tax Services. PwC employs over 125,000 people in more than 142 countries, *channeling knowledge and value through five lines of service and 22 industry-specialized practices.*

Careers

Actuaries at PwC work mainly in the Actuarial and Insurance Management Solutions group. They provide private and public organizations throughout the world with business insurance solutions.

Typical actuarial activities in casualty actuarial consulting involve a *broad range of risk analysis services related to personal and commercial lines, such as automobile liability, general liability and workers' compensation. Services include loss reserving, ratemaking, financial performance and strategy consulting, merger and acquisition valuations, reinsurance program review, and expert witness testimony.* The Casualty Actuarial Consulting group of PwC is the third largest casualty actuarial consulting organization, and the largest casualty actuarial consulting practice of any accounting firm in the United States.

Actuarial life insurance activities at PwC involve assisting clients *with critical strategic and/or financial planning issues, as well as operational or regulatory compliance aspects of life insurance companies. Services include financial analysis, taxation, litigation support, attestation, product development and many others.*

Actuaries at PwC are also active in insurance operations practice. They deliver a *broad range of claims and underwriting-related services to address insurance-related issues faced by a diverse clientele, including insurers, reinsurers, self-insureds, regulators, captives, capital markets and law firms. Services include claim and underwriting practices reviews, internal controls assessments, regulatory compliance diagnostics, market conduct examinations, litigation and due diligence support, claims portfolio valuations, and Managing General Agency controls reviews.*

Locations

In the United States alone, PwC has offices in Albany, Atlanta, Austin, Baltimore, Battle Creek, Birmingham, Bloomfield Hills, Boston, Buffalo, Cambridge, Century City, Charlotte, Chicago, Cincinnati, Cleveland, Columbus, Dallas, Dayton, Denver, Detroit, Florham Park, Fort Lauderdale, Fort Worth, Grand Rapids, Greensboro, Harrisburg, Hartford, Honolulu, Houston, Indianapolis, Irvine (Orange County), Jacksonville, Jersey City, Kansas City, Knoxville, Las Vegas, Lexington, Little Rock, Los Angeles, Louisville, McLean (Tysons Corner), Melville, Memphis, Menlo Park, Miami, Milwaukee, Minneapolis, Montgomery, Montpelier, New Haven, New Orleans, New York (HQ), Ogden, Orlando, Peoria, Philadelphia, Phoenix, Pittsburgh, Portland (Maine), Portland (Oregon), Raleigh, Richmond, Ridgewood, Rochester, Sacramento, Salt Lake City, San Diego, San Francisco, San Jose, Sarasota, Seattle, Spartanburg, St. Louis, Stamford, Syracuse, Tampa, Toledo, Tulsa, Washington, D.C., and West Palm Beach.

A.12 Towers Watson

- **Company Website:** www.towerswatson.com/
- **Careers:** www.towerswatson.com/careers/
- **Facebook:** www.facebook.com/towerswatsoncareers
- **LinkedIn:** www.linkedin.com/title/actuary/at-towers+watson/
- **Twitter:** twitter.com/towerswatson

In 2010, Towers Perrin and Watson Wyatt merged to become Towers Watson—a giant worldwide consulting firm headquartered in New York City. Today Towers Watson is a leading global professional services company that helps organizations improve performance through effective people, risk, and financial management. It has over 14,000 associates around the world and offers solutions in the areas of employee benefits, talent management, rewards, and risk and capital management.

Towers Perrin was one of the world's largest global management consulting firms, assisting organizations in managing people, performance and risk. It had provided innovative advice and assistance to large organizations in both the private and public sectors for more than 60 years. The firm's clients included three-quarters of the world's 500 largest companies and three-quarters of the Fortune 1000 largest U.S. companies. Towers Perrin had over 9,000 employees and 78 offices in 23 countries. Watson Wyatt Worldwide, on the other hand, was a global consulting firm focused on human capital and financial management. The company specialized in four areas: employee benefits, human capital strategies, technology solutions, and insurance and financial services. Watson Wyatt had more than 6,300 associates in 89 offices in 30 countries.

Careers

In its own words, *Towers Watson is an exciting place to launch your career. The worlds largest multinational corporations look to our talent to help them tackle their most pressing business issues. Youll have an opportunity to make your mark right here. Towers Watson welcomes the valuable perspectives and insights experienced professionals bring to the workplace.* The Towers Watson website presents numerous employee profiles, career paths, and personal histories and success stories that are both inspirational and practical.

Locations

The Towers Watson offices around the world are grouped into four regions: Asia Pacific, Europe Middle East and Africa, Latin America, and North America.

Asia/Pacific: Towers Watson has offices in Melbourne, Sydney, New South Wales (Australia); in Taipei (Taiwan); multiple offices in the People Republic of China: Beijing, Guangzhou, Shanghai, Shenzhen, Wuhan, Hong Kong SAR; in Mumbai, Bangalore, Delhi, and Kolkata (India); in Jakarta (Indonesia); in Tokyo (Japan); in Kuala Lumpur (Malasia); in Manila (Singapore); in Seoul (South Korea); in Bangkok (Thailand); and in Hanoi City and Ho Chi Minh City (Vietnam).

Europe/Middle East/Africa: Towers Watson has offices in Vienna (Austria), Brussels (Belgium), Paris (France), Cologne, Frankfurt, Munich, Reutlingen and Wiesbaden (Germany), Dublin (Ireland), Milan and Rome (Italy), Lisbon (Portugal), Moscow (Russian Federation), Cape Town and Johannesburg (South Africa), Madrid (Spain), Stockholm (Sweden), and Lausanne and Zurich (Switzerland).

Latin America: The Latin American offices of Towers Watson are in Buenos Aires (Argentina), Rio de Janeiro and São Paulo (Brazil), Santiago (Chile), Bogota (Columbia), Mexico City (Mexico), and Montevideo (Uruguay).

North America: The North America offices of Towers Watson are in Hamilton (Bermuda), Calgary, Montreal, Toronto, and Vancouver (Canada), and in the United States: Ann Arbor, Atlanta, Austin, Bloomington, Boston, Charlotte, Chicago, Cincinnati, Cleveland, Columbus, Dallas, Denver, Detroit, Grand Rapids, Hartford, Honolulu, Houston, Los Angeles, Memphis, Miami,

Milwaukee, Minneapolis, NY Tri-State Area, Philadelphia, Phoenix, Pittsburgh, Portland, San Antonio, San Diego, San Francisco, Seattle, St. Louis, Tampa, and Washington, D.C.

A.13 Other Firms

Many actuarial websites have up-to-date lists of actuarial consulting firms. For example, the website of the Canadian Institute of Actuaries maintains an extensive list at www.actuaries.ca.

Other lists may be found on the Society of Actuaries website, at the Career Center of the Casualty Actuarial Society, and by consulting many other agencies, both national and international. A look on the Internet for "actuarial consulting firms" will direct you to a plethora of consulting opportunities.

INSURANCE COMPANIES

An Internet search for the top ten insurance companies in the world led to the website www.popularsomething.com and produced the following ranking of these companies in 2009, based on their market share: American International Group (1), AXA Group (2), Allianz Worldwide (3), Manulife Financial (4), Generali Group (5), Prudential Financial (6), MetLife (7), Aviva (8), Munich Re Group (9), and AEGON (10).

All of these companies are profiled below. As in the case of the consulting firms profiled in Appendix A, the inclusion of other companies in this list of profiles is based on my personal knowledge of these companies acquired as director of the actuarial cooperative program at Concordia University and the objective of profiling both large and small companies.

All of these companies employ actuaries. The list is incomplete and somewhat random since there are simply too many companies to discuss in this guide. The directory of insurance companies on the Internet alone, found at www.iiin.com, lists thousands of companies, grouped into health, life, property and casualty, reinsurance, specialty insurance, and title insurance, with a large number of these companies represented internationally. In addition, most if not all of the large international banking institutions now have actuarial divisions. Only one or two of them are included in the section since the employment options that they provide tend to be quite similar.

As in the case of consulting firms, the companies profiled consist of many of the employers where the respondents to the survey on which this guide is based were working when they answered the survey or where they are working today. They include some of the major national and international employers of actuaries.

The descriptions that follow are merely meant to be starting points for your career search. As you browse through them, you should get a sense of what it is like to work as an actuary for different kinds of insurance companies. For this reason, the individual profiles stress different aspects of employment: global mobility, daily tasks, required qualifications, company philosophy, working conditions, and so on.

The quoted material can be found on the websites of the respective companies. You should consult these sites for more complete information. Job descriptions included with some company profiles do not indicate that these jobs are still available. They are meant to illustrate different aspects of actuarial life, and to give real-world examples of the main theme of this guide, which stresses the bond between mathematics, business, and statistics upon which actuarial careers are built.

Useful information about all of the companies below can also be found on the Facebook and LinkedIn websites of these companies. In the case of insurance companies, the postings on Twitter tend to be too personal and unpredictable to be included in this guide.

B.1 AEGON

- **Website:** www.aegon.com
- **Careers:** ca.indeed.com/Aegon-Canada-jobs
- **Facebook:** www.facebook.com/pages/Aegon/109569369069194
- **LinkedIn:** www.linkedin.com/company/aegon

AEGON is an international life insurance, pension, and asset management company with businesses in over 20 markets in the Americas, Europe, and Asia. AEGON employs some 26,500 people and serves 40 million customers worldwide. In 2008, the company developed its global talent management program aimed at developing leadership potential across the entire organization. The approach is designed to develop top performers, to build talent, and to make the company a preferred employer.

B.2 AIG

- **Website:** www.aig.com/
- **Careers:** www.aigcorporate.com/addresources/careers.html
- **Facebook:** www.facebook.com/pages/AIG/106038576102094
- **LinkedIn:** www.linkedin.com/company/aig

AIG believes that *progress is the result of the talent and dedication of thousands of employees around the globe, who have been steadfast in their commitment to serving customers and repaying taxpayers.* The company is a leading international insurance organization serving customers in more than 130 countries. The various AIG companies serve commercial, institutional, and individual customers through one of the most extensive worldwide property casualty networks of any insurer. The AIG companies are also leading providers of life insurance and retirement services in the United States.

B.3 | Allianz

- **Website:** https://www.allianz.com/en/index.html/
- **Careers:** https://www.allianz.com/en/careers/index.html
- **Facebook:**
  www.facebook.com/pages/Allianz-Life/108091675886108
- **LinkedIn:** www.linkedin.com/company/allianz-life

Allianz Group is an international financial services provider offering insurance, banking, asset management products and services to more than 76 million customers in about 70 countries. The company is one of the leading insurance groups in the world and ranks number one in the German property-casualty and life insurance markets. Allianz Group conducts business in almost every European country, with Germany, Italy, and France being its most important markets. In addition, the group has divisions in the United States, Central and Eastern Europe, and the Asia-Pacific region. Allianz Group is one of the five largest asset managers in the world.

B.4 | Aviva

- **Website:** www.avivacanada.com/
- **Careers:** www.aviva.com/careers/
- **Facebook:** www.facebook.com/Aviva
- **LinkedIn:** www.linkedin.com/company/aviva-life-insurance

Aviva is the world's seventh-largest insurance group and the biggest in the UK. It is one of the leading providers of life and pensions insurance in Europe and has substantial businesses elsewhere around the world. It is actively developing its long-term savings businesses in Asia Pacific and the United States. The company's main activities are long-term savings, fund management, and general insurance.

B.5   AXA

- **Website:** www.axa.com/
- **Careers:** www.axa.com/en/careers/
- **Facebook:** www.facebook.com/pages/AXA/112485985429503
- **LinkedIn:** www.linkedin.com/company/axa

In 1980, AXA did not exist. Since then, the Group has grown into a major international player. Today AXA is present in geographically diverse markets, with operations concentrated in Europe, North America, and Asia Pacific. In 2009, AXA was considered to have the second-highest market share in its field.

B.6   Desjardins

- **Website:** www.desjardins.com
- **Careers:** www.desjardinslifeinsurance.com/en/careers/Pages/careers.aspx
- **Facebook:** www.facebook.com/pages/Desjardins-Financial-Security/113735975303754
- **LinkedIn:** www.linkedin.com/company/desjardins-financial-security

Desjardins Financial Security has a customer base of more that 5 million Canadians offering tailor-made combinations of life and health insurance coverages. The company was formed by merging over 20 portfolios and companies. It pioneered the concept of bancassurance in Canada, and is a major player in the life and health insurance and financial services industry in the country. Today, Desjardins Financial Security ranks fourth among life and health insurers in Canada and first in Quebec in terms of written premiums.

B.7   Generali

- **Website:** www.generali-gw.com/
- **Careers:** www.generali-gi.com/careers/
- **Facebook:** www.facebook.com/pages/Assicurazioni-Generali/104036112966297
- **LinkedIn:** www.linkedin.com/company/generali

Generali Worldwide offers and administers employee benefits such as life and disability insurance, retirement and savings plans, and health insurance for companies and groups of all sizes, from small local companies to multinational companies and non-governmental organizations.

B.8   Great-West

- **Website:** www.greatwestlife.com/
- **Careers:** www.greatwestlife.com/001/Home/Careers/index.htm
- **Facebook:** www.facebook.com/pages/Great-West-Life/
  114741441875365
- **LinkedIn:** www.linkedin.com/company/great-west-healthcare

Great-West Life is a leading Canadian insurer. The company offers a broad portfolio of financial and benefit plan solutions for individuals, families, business owners, and organizations. A search for the top ten insurance companies in Canada ranked Great-West Life as the company with the greatest market share. This included Canada Life and London Life, two of its subsidiaries.

B.9   Manulife

- **Website:** www.manulife.ca/
- **Careers:** www.manulife.com/careers
- **Facebook:** www.facebook.com/pages/Manulife/110030142352890
- **LinkedIn:** www.linkedin.com/company/manulife-financial

Manulife has been doing business in Asia for more than 100 years. The company has one of the most extensive Asian operations of any life insurance company in the world with over 3,000 employees and almost 25,000 agents serving over 4 million customers. In April 2004, Manulife Financial Corporation merged with American financial services giant John Hancock Financial Services to make the largest life insurance company in North America.

B.10   MetLife

- **Website:** www.metlife.com/about/corporate-profile/index.html
- **Careers:** www.metlife.com/careers/index.html?WT.ac=GN_careers1
- **Facebook:** www.facebook.com/metlife
- **LinkedIn:** www.linkedin.com/company/metlife

MetLife is one of the largest global providers of insurance, annuities, and employee benefit programs, with 90 million customers in over 60 countries. The company states *that as a crucial part of our business and finance teams, actuaries are afforded a stimulating work environment, challenging assignments, and ample opportunity for professional development.*

B.11 | Munich Re

- **Website:** www.munichre.com/
- **Careers:** www.munichre.com/en/career/default.aspx
- **Facebook:** www.facebook.com/pages/Munich-Re/112684192080056
- **LinkedIn:** www.linkedin.com/company/munich-re

Munich Re is an international reinsurance company with more than 60 offices and subsidiaries worldwide. Mathematicians at Munich Re are considered "today's prophets." They work in life reinsurance, health reinsurance, non-life reinsurance, and information technology. Account managers at Munich Re serve as interfaces between Munich Re and the insurance companies. *As consultants for all matters related to life insurance, it is their duty to convey to the emerging markets the expe rience and knowledge that the company's specialists have acquired throughout the world. They deal with pricing, selecting and introducing new products, tax issues as well as the setting of terms and conditions, assessment of risks, distribution of insurance, and so on.*

B.12 | Prudential

- **Website:** www.prudential.com/
- **Careers:** www.prudential.com/view/page/public/12873
- **Facebook:** www.facebook.com/pages/Prudential-Financial/103763179663008
- **LinkedIn:** www.linkedin.com/company/prudential-insurance

Prudential Financial companies serve individual and institutional customers worldwide and include The Prudential Insurance Company of America, one of the largest life insurance companies in the United States ("The Rock"). These companies offer a variety of products and services, including life insurance, property and casualty insurance, mutual funds, annuities, pension and retirement related services and administration, asset management, securities brokerage, banking and trust services, real estate brokerage franchises, and relocation services. Prudential Financial has a global presence in the insurance industry.

B.13 | RSA Insurance Group

- **Website:** www.rsagroup.com/rsagroup/en/home
- **Careers:** www.rsagroup.ca/royalsun/en/default.asp?file=CareersOverview.htm
- **Facebook:** www.facebook.com/pages/RSA-Insurance-Group/106144822749281

RSA Insurance Group is one of the world's largest international insurance groups and employs approximately 50,000 individuals in over 50 countries. It is the second largest commercial and personal insurer in the United Kingdom. The three key values of RSA Insurance Group are truth, trust, and teamwork.

B.14 | Sun Life Financial

- **Website:** www.sunlife.ca/
- **Careers:** www.sunlife.ca/Canada/sunlifeCA/Careers
- **Facebook:** www.facebook.com/pages/Sun-Life/105825936123623
- **LinkedIn:** www.linkedin.com/company/sun-life-financial

Sun Life Financial is a leading financial services organization with operations in key markets around the world. The company offers a diverse range of financial products and services in wealth management and protection. *Wealth management includes asset management, mutual funds, pension plans and products, and annuities operations. Protection includes whole life, term life, universal life, unit-linked life and corporate-owned life insurance for individuals. As well, life, health and disability insurance products are offered to group clients.* Worldwide, Sun Life Financial has approximately 15,000 employees. In a recent survey, Sun Life was ranked as the second largest insurance company in Canada, based on market share.

B.15 | Swiss Re

- **Website:** www.swissre.com
- **Careers:** www.swissre.com/careers/
- **Facebook:** www.facebook.com/pages/Swiss Re/106010642764017
- **LinkedIn:** www.linkedin.com/company/swiss-re

Swiss Re sees reinsurance as an evolving industry and employs *innovative, forward-looking people who know the insurance industry's demands.* The company creates a flexible environment that promotes creativity. It *aims to be the pioneer that shapes reinsurance to reflect client requirements.* Swiss Reinsurance America Corporation is a division of Swiss Re, a worldwide reinsurance company with offices in Africa, Asia, Australasia, Europe, and North and South America.

B.16 | The Hartford

- **Website:** www.thehartford.com/
- **Careers:** www.thehartford.com/utility/careers/

- **Facebook:** www.facebook.com/TheHartford
- **LinkedIn:** www.linkedin.com/company/the-hartford

The Hartford has two divisions: Hartford Life and Hartford Property & Casualty. The company offers investment products, individual life insurance, group benefits, and property and casualty insurance. The company uses technology to enhance the quality of its services. It is a pioneer in the web-based delivery of financial services and is constantly updating its technology and service standards. *As a leader in insurance, asset management and financial service products, we offer professionals every possibility for growth. And whether we're helping customers or building careers, we're experts at creating the kind of advantages that help people reach their goals.*

RECIPROCITY AGREEMENTS

The Institute and Faculty of Actuaries in the United Kingdom has signed mutual recognition agreements with several actuarial organizations: the Institute of Actuaries of Australia, the Canadian Institute of Actuaries, the Society of Actuaries of the United States, the Institute of Actuaries of Japan, and the *Groupe Consultatif* of the European Union.

The agreement describes the process, country by country, by which actuaries in the countries involved can become members of the actuarial societies in the other participating countries. The document, summarized in this appendix, is the official description of the reciprocity agreement and should be consulted for specific details.

The agreement says, in essence, that actuaries who have become Fellows of a national actuarial society by the normal route (having passed the necessary examinations), and who are members in good standing (having paid the annual membership fee in their home country), meet the professionalism requirements of the guest country, fulfill the necessary residency requirements, and intend to practice in the guest country, can do so by reciprocity. Anyone interested in taking advantage of this agreement should consult the relevant documents published on the Institute and Faculty of Actuaries and *Groupe Consultatif* websites:

www.actuaries.org.uk/members/pages/mutual-recognition-qualifications

www.gcactuaries.org/documents/MRA_FINAL_010111.pdf

C.1   United Kingdom and Australia

Fellows of the Institute of Actuaries of Australia in good standing can become Fellows of the Institute and Faculty of Actuaries on the following conditions:

- They have become full members of the IAAust by examination and not in recognition of membership of another actuarial association.

- They wish to become practicing actuaries in the United Kingdom or Republic of Ireland or intend to advise on UK or Irish business.

- They have at least three years' recent appropriate practical experience of which at least one year was of UK or the Republic of Ireland business.

- They have attended an approved professionalism course.

Conversely, the Institute of Actuaries of Australia will admit to Accredited Member status of the IAAust Fellows of the Institute and Faculty of Actuaries who wish to pursue actively the profession of actuary in Australia, provided that they satisfy the following conditions:

- They have qualified as Fellows of the Institute and Faculty of Actuaries through examination.

- They have been a resident in Australia for at least 6 months.

- They gained suitable experience in local actuarial practice.

- They have passed a recognized professionalism course within the previous 5 years or earlier at the discretion of the Committee, or any other course approved by the Committee.

Applicants who meet these conditions will automatically be recommended to the Council of the Institute of Actuaries of Australia for membership.

C.2   United Kingdom and Canada

Canadian actuaries intending to practice in the United Kingdom or Republic of Ireland can become Fellows of the Institute and Faculty of Actuaries on conditions similar to those for Australian actuaries, provided that they are full members of the Canadian Institute of Actuaries through examination from the Society of Actuaries, the Casualty Actuarial Society, or the Institute of Actuaries of Australia.

The Canadian Institute of Actuaries will, in turn, admit as a permanent Member of the Canadian Institute of Actuaries, a Fellow of the Institute and Faculty of Actuaries, who wishes to pursue actively the profession of actuary in Canada, on conditions similar to the Australian case. Applicants must also have passed the Practice Education Course (PEC) of the Canadian Institute of Actuaries. As stipulated in the reciprocity agreement, this course *may be attended following twelve months of relevant Canadian experience.* Candidates must have completed at least twelve hours of Canadian Institute of Actuaries-approved professional development (PD) in the twelve months prior to the application for Membership status. Candidates must also demonstrate that they have completed three years of practical actuarial work within the three-year period immediately prior to their application for Membership.

C.3 | United Kingdom and the United States

The agreement between the Institute and Faculty of Actuaries and the Society of Actuaries is similar to the two previous examples. To become an accredited member of the Society of Actuaries, an applicant must fulfill the following conditions:

- Have attained full membership of the Institute and Faculty of Actuaries by examination and not in recognition of membership of another actuarial association.

- Be a Fellow in good standing of the Canadian Institute of Actuaries, or Member in good standing of the American Academy of Actuaries, or full member in good standing of other actuarial associations designated from time to time by the Society of Actuaries Board of Governors.

- Have attended and passed the Society of Actuaries Fellowship Admissions Course, or its equivalent as recognized by the Society of Actuaries, in the five years prior to application.

- Have satisfied the Society of Actuaries Professional Development requirements, or its equivalent as recognized by the Society of Actuaries, in the five years prior to application.

C.4 | United Kingdom and Japan

The Mutual Recognition Agreement between the Institute of Actuaries (UK) and the Institute of Actuaries (Japan), released in 2011, provides a reciprocal way for Fellows of the UK and Japan Institutes to participate in the activities of the other organization. The agreement is helpful for Fellows of the Japanese Institute whose companies have operations in the UK, and for Fellows of the UK Institute who are posted in Japan for a limited period of time. It is based on an earlier agreement dating back to the year 2000.

The agreement states that *the IAJ agrees to admit to its 'Kenkyu-Kaiin' status any Fellow of the IoA who submits an application form and pays the required fee, subject to any conditions prescribed for such status.* The agreement states further that *Fellows of either organization taking advantage of the stated program of the other organization will receive all mailings, including publications and meeting notices, except for mailings relating to matters requiring a membership vote. They may attend all meetings and programs of the other organization at the same rate charged to members, although they may be excluded from business meetings at which membership votes are to be taken if this is considered desirable by the organization scheduling the vote.*

C.5   United Kingdom and the European Union

The Institute of Actuaries (UK) is a signatory to the *Groupe Consultatif* Agreement on mutual recognition of qualifications. The details of the agreement can be found on the following website:

     http://www.actuaries.org.uk/research-and-resources/
documents/mutual-recognition-agreements-groupe-consultatif

The agreement states that *each association shall designate the class or classes of members of that association which are to be regarded as full members for the purpose of this Agreement, and shall maintain a list of such members.*

The conditions for membership require that each member association of the *Groupe Consultatif* maintain and provide to the group a list of actuaries who are members of other associations of the *Groupe*. Essentially, membership is granted without requiring further training, the passing of examinations or periods of experience. However, conditions apply. For example, it follows from the code of conduct of individual associations that professional actuarial services in a host country should only be undertaken if the applicant has the appropriate experience to do so.

C.6   Within the European Union

The *Groupe Consultatif Actuariel Européen* has established reciprocity agreements for the recognition of actuarial qualification between the following national associations of actuaries in Europe: the Austrian Actuarial Society, Royal Association of Belgian Actuaries, Bulgarian Actuarial Society, the Cyprus Association of Actuaries, the Czech Society of Actuaries, the Danish Society of Actuaries, the Estonian Actuarial Society, the Actuarial Society of Finland, the Institute of Actuaries of France, the German Actuarial Society, the Hellenic Actuarial Society (Greece), the Hungarian Actuarial Society, the Society of Icelandic Actuaries, the Society of Actuaries in Ireland, the National Council of Actuaries (Italy), the Italian Institute of Actuaries, the Latvian Actuarial Association, the Lithuanian Actuarial Society, the Luxembourg Actuarial Association, the Actuarial Society of the Netherlands, the Norwegian Society of Actuaries, the Polish Society of Actuaries, the Portuguese Institute of Actuaries, the Slovak Society of Actuaries, the Slovenian Association of Actuaries, the Spanish Institute of Actuaries, the Catalan Actuarial Association (Spain), the Swedish Society of Actuaries, the

Swiss Association of Actuaries, and the Institute and Faculty of Actuaries (United Kingdom). Essentially, the agreement means that full-fledged actuaries have professional mobility throughout Europe. In addition, the following European societies have observer status: the Channel Islands Actuarial Society, the Croatian Actuarial Association, the Romanian Actuarial Association, the Actuarial Society of Turkey, and the Society of Actuaries of Ukraine.

ACTUARIAL WEBSITES

D.1   North American Sites

1. Actuarial Foundation (`www.actuarialfoundation.org`)

2. American Academy of Actuaries (`www.actuary.org`)

3. American Society of Pension Professionals & Actuaries (`www.asppa.org`)

4. Canadian Institute of Actuaries (`www.actuaries.ca`)

5. Casualty Actuarial Society (`www.casact.org`)

6. Conference of Consulting Actuaries (`www.ccactuaries.com`)

7. Joint Board for Consulting Actuaries (`www.irs.gov/taxpros/actuaries/index.html`)

8. Society of Actuaries (`www.soa.org`)

D.2   International Sites

1. Aktuarvereinigung Österreichs (Austria) (`www.avoe.at`)

2. Asociacion Mexicana de Actuarios (Mexico) (`www.amac.org.mx`)

3. Association Suisse des Actuaires (Switzerland) (`www.actuaries.ch`)

4. Actuarial Society of Hong Kong (`www.actuaries.org.hk`)

5. Actuarial Society of India (`www.actuariesindia.org`)

6. Actuarial Society of Malaysia (`www.actuaries.org.my`)

7. Actuarial Society of South Africa (`www.actuarialsociety.org.za/`)

8. Den Danske Aktuarforening (Denmark) (`www.aktuarforeningen.dk`)

9. Deutsche Aktuarvereinigung (Germany) (`www.aktuar.de`)

10. Den Norske Aktuarforening (Norway) (`www.aktfor.no`)

11. Groupe Consultatif Actuariel Europeen (European Union) (`www.gcactuaries.org`)

12. Het Actuarieel Genootschap (Netherlands) (`www.ag-ai.nl`)

13. Institute of Actuaries of Australia Actuarial Society (`http://www.actuaries.asn.au/`)

14. Institute and Faculty of Actuaries (UK) (`www.actuaries.org.uk`)

15. Institut des Actuaires Français (France) (`www.institutdesactuaires.com`)

16. Institute of Actuaries of Japan (`www.actuaries.jp/english/index.html`)

17. Instituto Brasileiro de Atuária (Brazil) (`www.atuarios.org.br/iba/`)

18. Instituto de Actuarios Españoles (Spain) (`www.actuarios.org`)

19. International Actuarial Association (`www.actuaries.org`)

20. International Association of Consulting Actuaries (`www.actuaries.org/index.cfm?DSP=IACA&ACT=INDEX&lang=EN`)

21. Israel Association of Actuaries (`www.actuaries.org.il`)

22. Istituto Italiano degli Attuari (`www.italian-actuaries.org`)

23. Japanese Society of Certified Pension Actuaries (`www.jscpa.or.jp`)

24. New Zealand Society of Actuaries (`www.actuaries.org.nz`)

25. Portuguese Institute of Actuaries (`www.actuarial.pt`)

26. Society of Actuaries in Ireland (`web.actuaries.ie/`)

27. Suomen Aktuaariyhdistys (Finland) (`www.actuary.fi`)

28. Svenska Aktuarieföreningen (Sweden) (`www.aktuarieforeningen.se/`)

29. The Chinese Actuarial Club (China) (`www.chinese-actuary.org`)

RECRUITING AGENCIES

E.1 | Prominent Agencies

The recruiting agencies listed below form a small set of the large number of well-known companies in the United States, Canada, UK, and elsewhere that can serve as a starting point for you in your search for an ideal actuarial position worldwide. You can get information into the status of these companies and others by consulting *Dun and Bradstreet* at www.dnb.com, the world's leading source of commercial information and insight on businesses. In addition, you will find job listings using actuary-specific search engines such as www.job4actuary.com

- ACTEX Actuarial Recruiting (Winsted, CT) (www.searchfirm.com/snapshot/ACTEX_Actuarial_Recruiting_Winsted.asp) *Combining career counseling and placement expertise with the strength and resources of ACTEX Publications, ACTEX Actuarial Recruiting is positioned to provide effective career counseling and to meet clients' needs for top quality actuarial talent.*

- Actuarial Careers (White Plains, NY) (www.actuarialcareers.com) *Exclusively dedicated to the placement and advancement of Chief Actuaries, Fellows, Associates, and Student Actuaries.*

- Acumen Resources (London, UK) (www.acumen-resources.com) *Acumen Resources is one of the world's leading actuarial recruitment agencies. All our consultants have either actuarial or human resources qualifications and have worked in a wide variety of practice areas. This allows you to tap into a significant global network and maximize your chances of finding the right actuarial job with the right company.*

- Andover Research (New York, NY) (`www.andoverresearch.com`) "Andover Research Ltd. specializes in the recruitment and placement of actuaries and benefit consultants worldwide."

- D. W. Simpson and Company (Chicago, IL) (`www.dwsimpson.com`) *DW Simpson Global Actuarial Recruitment serves the actuarial profession worldwide in all disciplines, recruiting at all levels from Entry-Level through Fellowship, and works with clients on both retained and contingent searches.* The company recruits actuaries for employment in casualty, health, life, and pension for Canada, the U.S., the UK, Europe, India, Asia, Australia, the Middle East, and other regions.

- Darwin Rhodes (London, UK) (`www.darwinrhodes.com`) *Darwin Rhodes has its headquarters in the City of London with specialist teams covering the UK market.* The company covers the actuarial and insurance markets of mainland Europe and operates from offices in both London and Zurich, Switzerland. Its offices in Hong Kong and Shanghai cover the major financial centers of East and Southeast Asia including Tokyo and Singapore and emerging centers such as Kuala Lumpur, Taipei, Bangkok, and Ho Chi Minh City.

  Darwin Rhodes was the first insurance recruiter to enter the Indian market and its New York and Chicago offices cover the major North American financial centers as well as developing markets in Central and South America. Its Sydney office covers both Australia and New Zealand.

- Elliot Bauer (London, UK) (`www.elliottbauer.com`) Elliott Bauer is a leading actuarial recruitment firm. It prides itself in finding the best candidates for major firms around the world.

- Emerald Group (London, UK) (`www.emerald-group.com`) The Emerald Group originated in the UK as an organization serving the actuarial profession at all levels. It provides staff to many of the worlds largest consultancy and Investment houses across five continents.

- Mid America Search (West Des Moines, IA) (`www.midamericasearch.com`) *Mid America Search is an executive search firm specializing primarily in the insurance and financial industry. It conducts personnel searches for hundreds of outstanding insurance companies and related organizations both in the United States and internationally.*

- Pinnacle Group (Portsmouth, NH) (`www.pinnaclejobs.com`) The Pinnacle Group specializes in actuarial recruitment. Its clients include leading insurance, consulting and investment firms in the United States and elsewhere. Its areas of recruitment include life/annuity, health/managed care, property and casualty, pension and benefits, and reinsurance.

- Pryor Associates (Hicksville, NY) (`www.ppryor.com`) Pryor Associates agency provides employees for all facets of the insurance industry including administration, underwriting, claims, loss control, accounting, actuarial, pension, property/casualty, life/health, employee benefits, agency/brokerage, and insurance/reinsurance company/corporate placements.

- SC International (Downers Grove, IL) (`www.scinternational.com`) International focuses primarily on the actuarial, employee benefits, and insurance market. Like many of the actuarial recruitment agencies, the company plays *an integral role in both sourcing personnel and coordinating the interview process, from initial conversations to meetings and final negotiations.*

- Stewart Search Advisors (Portsmouth, NH) (`www.stewartsearch.com`) and (`www.actuarialfutures.com`) Stewart Search specializes in the placement of actuaries in the areas of life, health, pension, property & casualty, reinsurance, consulting, investment, and banking. Its clients range in size from smaller specialty firms to Fortune 100 companies.

E.2 Sample Job Postings

You might find it interesting to see samples of actual job postings for actuarial positions as described on the career section of the SOA website. You can search this website by keywords, job function, industry, and state (including international and non-U.S. positions).

Valuation Actuary

A search for "valuation actuary" produced 30 hits. Here is a sample:

Position Title:	Actuarial Analyst
Job Type:	Full-Time
Job Duration:	Indefinite
Industry:	Life
Minimum Education:	BA/BS/Undergraduate
Job Function:	Actuarial Candidate: 3–5 Exams
Minimum Experience:	3–5 Years
Entry Level:	No
Required Travel:	0–10%
Location:	Connecticut

Pricing Actuary

A search for "pricing actuary" produced 62 hits. Here is a sample:

Position Title:	Actuary (International Pricing)
Job Type:	Full-Time
Job Duration:	Indefinite
Industry:	Insurance
Minimum Education:	BA/BS/Undergraduate
Job Function:	FSA
Minimum Experience:	Over 10 Years
Entry Level:	No
Required Travel:	0–10%
Location:	Delaware

Consulting Actuary

A search for "consulting actuary" produced 125 hits. Here is a sample:

Position Title:	P & C Actuary
Job Type:	Full-Time
Job Duration:	Indefinite
Industry:	Property/Casualty
Minimum Education:	BA/BS/Undergraduate
Job Function:	International (ACAS with 5+ Exams, Auto Pricing Experience a Plus)
Minimum Experience:	5–7 Years
Entry Level:	No
Required Travel:	None
Location:	British Columbia, Manitoba, Newfoundland (Canada)

Pension Actuary

A search for "pension actuary" produced 117 hits. Here is a sample:

Position Title:	Life Analyst
Job Type:	Full-Time
Job Duration:	Indefinite
Industry:	Life
Minimum Education:	BA/BS/Undergraduate

Job Function:	ASA
Minimum Experience:	3–5 Years
Entry Level:	No
Required Travel:	None
Location:	International: Australia, United Kingdom

Financial Actuary

A search for "financial actuary" produced 150 hits. Here is a sample:

Position Title:	P & C Actuary
Job Type:	Full-Time
Job Duration:	Indefinite
Industry:	Life
Minimum Education:	BA/BS/Undergraduate
Job Function:	International (ACA with 5–7 Years Non-Life Insurance)
Minimum Experience:	5–7 Years
Entry Level:	No
Required Travel:	None
Location:	International: Thailand, Taiwan

What all of these positions have in common is that in each case, applicants must have at least three years of experience. What is also interesting is that none of the candidates is required to have a university degree higher than a bachelor's degree. In all cases, experience counts. If you peruse many of the other advertised positions, you will find that this is quite common to most positions. Moreover, very few of the advertised positions require extensive travel, no matter where the location. What is significant is that even such a relatively casual search produces hundreds of job openings. It again shows that the actuarial profession provides one of the most desirable and secure career choices worldwide.

SOA EDUCATION SUMMARY

F.1 Associateship Requirements

Validation of Educational Experience

- VEE Economics
- VEE Corporate Finance
- VEE Applied Statistics

Examinations

- Exam P: Probability
- Exam FM: Financial Mathematics
- Exam MFE: Models for Financial Economics
- Exam MLC: Models for Life Contingencies
- Exam C: Construction and Evaluation of Actuarial Models

Fundamentals of Actuarial Practice (FAP) eLearning Course

Associateship Professionalism Course (APC)

F.2 Fellowship Requirements

In addition to the Associateship requirements, a Fellow must complete the modules and courses in one of the following five tracks, and must pass the **Decision**

Making and Communication (DMAC) Module and the **Fellowship Admissions Course (FAC):**

1. Finance/ERM Track

- Advanced Finance/ERM Exam
- Financial Economic Theory and Engineering Exam
- Financial and Health Economics Module
- Financial Reporting Module
- Operational Risk Module

2. Investment Track

- Advanced Portfolio Management Exam
- Financial Economic Theory and Engineering Exam
- Financial and Health Economics Module
- Investment Strategy Module
- Operational Risk Module or Financial Reporting Module

3. Individual Life and Annuities Track

- Individual Life and Annuities Company/Sponsor Perspective (CSP) Exam
- Individual Life and Annuities Design and Pricing (DP) Exam
- Financial and Health Economics Module
- Regulation and Taxation Module
- Operational Risk Module or Financial Reporting Module

4. Retirement Benefits Track

- Retirement Benefits Company/Sponsor Perspective (CSP) Exam
- Retirement Benefits Design and Pricing (DP) Exam
- Financial and Health Economics Module
- Social Insurance Module
- Operational Risk Module or Investment Strategy Module
- Enrolled Actuaries (EA) Exams (U.S. Only)

5. Group and Health Track

- Group and Health Company/Sponsor Perspective (CSP) Exam
- Group and Health Design and Pricing (DP) Exam
- Financial and Health Economics Module

- Health Foundations Module
- Pricing, Reserving and Forecasting Module

F.3 Chartered Enterprise Risk Analyst (CERA) Requirements

Validation of Educational Experience

- VEE Economics

Examinations

- Exam P: Probability
- Exam FM: Financial Mathematics
- Exam MFE: Models for Financial Economics
- Exam C: Construction and Evaluation of Actuarial Models
- Advanced Finance/ERM Exam

Modules

- Operational Risk Module

Associateship Professionalism Course (APC)

CAS EDUCATION SUMMARY

G.1 Associateship Requirements

Validation by Educational Experience

- VEE Applied Statistical Methods
- VEE Corporate Finance
- VEE Economics

Online Courses

- Online Course 1: Risk Management and Insurance Operations (same as Course CA1 of the Institute of Actuarries (UK))
- Online Course 2: Insurance Accounting, Coverage Analysis, Insurance Law and Insurance Regulation (same as Course CA2 of the Institute of Actuarries (UK))

Examinations

- Exam 1: Probability (same as SOA Exam P)
- Exam 2: Financial Mathematics (same as SOA Exam FM)
- Exam 3: Actuarial Models: Segment 3F, Financial Economics (same as SOA Exam MFE) and Segment 3L, Life Contingencies and Statistics
- Exam 4: Construction and Evaluation of Actuarial Models (same as SOA Exam C)

- Exam 5: Basic Techniques for Ratemaking and Estimating Claim Liabilities
- Exam 6: Regulation and Financial Reporting (Nation-Specific)

Course on Professionalism

G.2 Fellowship Requirements

Examinations

- Exam 7: Estimation of Policy Liabilities, Insurance Company Valuation, and ERM
- Exam 8: Advanced Ratemaking
- Exam 9: Financial Risk and Rate of Return

For more details on the preliminary examinations 1/P, 2/FM, 3F/MFE, and 4/C of the Canadian Institute of Actuaries, the Casualty Actuarial Society, and the Society of Actuaries, see the website www.actuarialexams.org/ preliminaryexam. The Institute administers the exams for the Online Courses 1/CA1 and 2/CA2.

ACTUARIAL SYMBOLS

True to the origin of their name, actuaries use an extensive list of special symbols for their work. It's a kind of cleverly devised shorthand for actuarial objects and functions. The notation is based on principles for construction adopted by the Second International Congress of Actuaries in London in 1898. The list is modified and updated from time to time with the approval of the Permanent Committee of Actuarial Notations of the International Actuarial Association. Appendices 3 and 4 of the standard reference on actuarial mathematics by Bowers et al. (see [5]) contains a full list of symbols. The following list gives an indication of the kind of symbols involved. Many of these symbols occur in the sample questions and answers in Chapter 2. For missing symbols used in these examples, we refer to [5] for their definition and explanation.

- The symbol $a_{\overline{n}|}$ denotes the value of an annuity of one dollar per year for n years, payable at the end of each year.
- The symbol $\ddot{a}_{\overline{n}|}$ denotes of value of an annuity of one dollar per year for n years, payable at the beginning of each year.
- The symbol $a_{\overline{n}|i}$ denotes the value of an annuity of one dollar per year for n years at i percent interest per year, payable at the end of each year.
- The symbol $\ddot{a}_{\overline{n}|i}$ denotes the value of an annuity of one dollar per year for n years at i percent interest per year, payable at the beginning of each year.
- The symbol $\overline{a}$ denotes an annuity payable continuously at the rate of one dollar per year.

- The symbol $\ddot{a}_{xy}$ denotes an annuity payable during the joint lives of (x) and (y), payable at the beginning of each year.
- The symbol $\ddot{a}_{\overline{xy}}$ denotes an annuity payable as long as one of (x) and (y) is alive, payable at the beginning of each year.
- The symbol $\ddot{a}_{x:\overline{n}|}$ denotes an annuity payable during the joint duration of the life of (x) and a term of n years.
- The symbol $_nd_x$ denotes the expected number of deaths between the ages x and $x+n$.
- The symbol $\overset{\circ}{e}_x$ denotes the average remaining lifetime at age x.
- The symbol l_x denotes the expected number of survivors to age x from l_0 newborns.
- The symbol p_x denotes the probability that life (x) will reach age $x+1$.
- The symbol $_tp_x$ denotes the probability that life (x) will survive the next t years.
- The symbol P_x denotes the annual premium of a whole life policy of one dollar, payable upon the death of x, with the first premium payable when the policy is issued.
- The symbol P_{xy} denotes the annual premium of a whole life policy of one dollar, payable upon the death of x, with the first premium payable when the policy is issued.
- The symbol $P_{\overline{xy}}$ denotes the annual premium of a whole life policy of one dollar, payable upon the first death of x or y, with the first premium payable when the policy is issued.
- The symbol q_x denotes the probability that life (x) will die within the next year.
- The symbol $_tq_x$ denotes the probability that life (x) will die within t years.
- The symbol $s(x)$ denotes the probability that a newborn will reach age x.
- The symbol (x) denotes a living person age x.
- The symbol (xy) denotes two living persons age x and y, respectively.

BIBLIOGRAPHY

[1] *Actuarial Career Planner*. The Society of Actuaries, Schaumburg, Illinois, 1998.

[2] Alexander, D., *Steps Forward in China*. International Section Newsletter, Number 24, February 2001, The Society of Actuaries, Schaumburg, Illinois.

[3] *Basic Education Catalog* (Fall 2002). The Society of Actuaries, Schaumburg, Illinois, 2002.

[4] *Basic Education Catalog* (Spring 2003). The Society of Actuaries, Schaumburg, Illinois, 2003.

[5] Bowers, N. L., Gerber, H. U., Hickman, J. C., Jones, D. A., and Nesbitt, C. J., *Actuarial Mathematics* (2nd Edition). The Society of Actuaries, Schaumburg, Illinois, 1997.

[6] Brown, R. L., *The Globalisation of Actuarial Education*. British Actuarial Journal, Volume 8, Number 1, 2002, pages 1–3.

[7] *Be An Actuary*. www.beanactuary.org, 2011.

[8] *CAS Survey on Professional Skills*. The Casualty Actuarial Society, Arlington, Virginia, 2002.

[9] *CAS Syllabus*. The Casualty Actuarial Society, Arlington, Virginia, 2002.

[10] *Encyclopaedia Britannica* (11th Edition). Volume XXII, London, 1911.

[11] Institute and Faculty of Actuaries, *Actuaries in Risk Management.* www
.actuaries.org.uk, 2011.

[12] Jones, B. L., *An Introduction to Actuarial Models and Modeling (An Interactive Approach).* Actex Publications, Winsted, Connecticut, 2000.

[13] Kellison, S. G., *The Theory of Interest* (2nd Edition). Irwin McGraw-Hill,
Boston, 1991.

[14] Klugman, S. A., Panjer, H. H., and Willmot, G. E., *Loss Models: From Data
to Decisions.* John Wiley and Sons, New York, 1998.

[15] Krantz, L., and Lee, T., *Jobs Rated Almanac 2002* (6th Edition). Barricade,
2002.

[16] Morgan, E., *Love it or hate it.* The Actuary, Staple Inn Actuarial Society,
July 2001.

[17] Ott, R. L., and Longnecker, M., *Statistical Methods and Data Analysis* (5th
Edition). Duxbury, Pacific Grove, California, 2001.

[18] Perryman, F. S., *International Actuarial Notation.* Proceedings of the Casualty Actuarial Society, Volume 36, Number 66, 1949, pages 123–131.

[19] Szabo, F. E., *Graduate Student Success.* Harcourt Brace Canada, Toronto,
Ontario, 1995.

[20] Szabo, F. E., *Linear Algebra: An Introduction Using Maple.* Harcourt/
Academic Press, Boston, Massachusetts, 2002.

[21] Szabo, F. E., *Actuaries' Survival Guide.* (1st Edition). Elsevier Academic
Press, Boston, Massachusetts, 2004.

[22] *The Future Actuary.* The Society of Actuaries, Schaumburg, Illinois, 2011.

[23] Wackerly, D. D., Medenhall, W., and Scheaffer, R. L., *Mathematical
Statistics with Applications* (5th Edition). Wadsworth Publishing Company,
Belmont, California, 1996.

[24] Warren, W. S., *The Physical Basis of Chemistry* (2nd Edition). Harcourt/
Academic Press, Boston, 2000.

[25] Zima, P., and Brown, R. L., *Mathematics of Finance* (5th Edition). McGraw-
Hill Ryseron, Toronto, 2001.

INDEX